Communications in Computer and Information Science 2719

Series Editors

Gang Li, *School of Information Technology, Deakin University, Burwood, VIC, Australia*

Joaquim Filipe, *Polytechnic Institute of Setúbal, Setúbal, Portugal*

Zhiwei Xu, *Chinese Academy of Sciences, Beijing, China*

Rationale
The CCIS series is devoted to the publication of proceedings of computer science conferences. Its aim is to efficiently disseminate original research results in informatics in printed and electronic form. While the focus is on publication of peer-reviewed full papers presenting mature work, inclusion of reviewed short papers reporting on work in progress is welcome, too. Besides globally relevant meetings with internationally representative program committees guaranteeing a strict peer-reviewing and paper selection process, conferences run by societies or of high regional or national relevance are also considered for publication.

Topics
The topical scope of CCIS spans the entire spectrum of informatics ranging from foundational topics in the theory of computing to information and communications science and technology and a broad variety of interdisciplinary application fields.

Information for Volume Editors and Authors
Publication in CCIS is free of charge. No royalties are paid, however, we offer registered conference participants temporary free access to the online version of the conference proceedings on SpringerLink (http://link.springer.com) by means of an http referrer from the conference website and/or a number of complimentary printed copies, as specified in the official acceptance email of the event.

CCIS proceedings can be published in time for distribution at conferences or as post-proceedings, and delivered in the form of printed books and/or electronically as USBs and/or e-content licenses for accessing proceedings at SpringerLink. Furthermore, CCIS proceedings are included in the CCIS electronic book series hosted in the SpringerLink digital library at http://link.springer.com/bookseries/7899. Conferences publishing in CCIS are allowed to use our online conference service (Meteor) for managing the whole proceedings lifecycle (from submission and reviewing to preparing for publication) free of charge.

Publication process
The language of publication is exclusively English. Authors publishing in CCIS have to sign the Springer CCIS copyright transfer form, however, they are free to use their material published in CCIS for substantially changed, more elaborate subsequent publications elsewhere. For the preparation of the camera-ready papers/files, authors have to strictly adhere to the Springer CCIS Authors' Instructions and are strongly encouraged to use the CCIS LaTeX style files or templates.

Abstracting/Indexing
CCIS is abstracted/indexed in DBLP, Google Scholar, EI-Compendex, Mathematical Reviews, SCImago, Scopus. CCIS volumes are also submitted for the inclusion in ISI Proceedings.

How to start
To start the evaluation of your proposal for inclusion in the CCIS series, please send an e-mail to ccis@springer.com

Mirko Presser · Antonio Skarmeta · Srdjan Krco
Editors

Global Internet of Things and Edge Computing Summit

Second International Summit, GIECS 2025
Madrid, Spain, September 22, 2025
Proceedings

Springer

Editors
Mirko Presser
Aarhus University
Herning, Denmark

Antonio Skarmeta
University of Murcia
Murcia, Spain

Srdjan Krco
DunavNET
Novi Sad, Serbia

ISSN 1865-0929 ISSN 1865-0937 (electronic)
Communications in Computer and Information Science
ISBN 978-3-032-09554-1 ISBN 978-3-032-09555-8 (eBook)
https://doi.org/10.1007/978-3-032-09555-8

This work was supported by Alliance for AI, IoT and Edge Continuum.

© The Editor(s) (if applicable) and The Author(s) 2026. This book is an open access publication.

Open Access This book is licensed under the terms of the Creative Commons Attribution 4.0 International License (http://creativecommons.org/licenses/by/4.0/), which permits use, sharing, adaptation, distribution and reproduction in any medium or format, as long as you give appropriate credit to the original author(s) and the source, provide a link to the Creative Commons license and indicate if changes were made.
The images or other third party material in this book are included in the book's Creative Commons license, unless indicated otherwise in a credit line to the material. If material is not included in the book's Creative Commons license and your intended use is not permitted by statutory regulation or exceeds the permitted use, you will need to obtain permission directly from the copyright holder.
The use of general descriptive names, registered names, trademarks, service marks, etc. in this publication does not imply, even in the absence of a specific statement, that such names are exempt from the relevant protective laws and regulations and therefore free for general use.
The publisher, the authors and the editors are safe to assume that the advice and information in this book are believed to be true and accurate at the date of publication. Neither the publisher nor the authors or the editors give a warranty, expressed or implied, with respect to the material contained herein or for any errors or omissions that may have been made. The publisher remains neutral with regard to jurisdictional claims in published maps and institutional affiliations.

This Springer imprint is published by the registered company Springer Nature Switzerland AG
The registered company address is: Gewerbestrasse 11, 6330 Cham, Switzerland

If disposing of this product, please recycle the paper.

Preface

This volume presents the select proceedings of the Global IoT and Edge Computing Summit (GIECS) 2025. Its content will be of interest to academia and industry professionals interested in advancing knowledge and practice in computation, communication, and engineering.

GIECS is an international forum established to showcase the latest research findings and foster dialogue in IoT and Edge Computing. Through a systematic peer review process, contributions from researchers, developers, and practitioners at the intersection of cloud, edge, and IoT were evaluated. The most significant papers were selected to foster a deeper understanding of how continuum computing is shaping the future of the Internet.

The conference invited submissions on a range of topics, including but not limited to:

- IoT Enabling Technologies
- IoT Applications, Services, and Real-World Implementations
- End-user and Human-Centric IoT, including IoT Multimedia, Societal Impacts, and Sustainable Development
- IoT Security, Privacy, and Data Protection
- IoT Pilots, Testbeds, and Experimental Results

From 24 submissions, 14 full papers were selected through a rigorous single-blind review. Each paper underwent two review rounds, with feedback provided by three independent domain experts. This process ensured that only high-quality research was included.

Handling of Committee Members' Contributions

In maintaining the highest standards of academic integrity and impartiality, we ensured that contributions co-authored by members of the GIECS 2025 committees were reviewed with strict adherence to a transparent and objective process.

Of the 14 accepted papers, six were co-authored by committee members. To prevent any conflict of interest, the following measures were implemented:

- **Blind Review Process:** All submissions, including those co-authored by committee members, underwent a single-blind review in which the identities of the reviewers were unknown to the authors and the committee members. This ensured that papers were evaluated and selected solely on their scholarly merit.
- **Independent Reviewers:** Committee members who were co-authors of any submission were entirely excluded from the selection of reviewers. The reviewers selected to review these submissions were independent domain experts unaffiliated with the authors.

- **Score-based Selection of Submissions**: The acceptance of submissions was determined by a total score from the three independent reviews of each paper. Accordingly, all accepted submissions were objectively selected based on the highest scores, with committee members who co-authored any submission fully excluded from the decision-making process for their papers.

By implementing these measures, we ensured that all accepted papers, regardless of authorship, were judged fairly and in accordance with the conference's review criteria. Thus, the papers featured in this volume were selected based on the quality and impact of their research, without any bias or preferential treatment.

GIECS is held annually in collaboration with the Alliance for IoT and Edge Computing Innovation (AIOTI) established by the European Commission. The 2025 edition took place in Madrid, Spain, and featured keynote speeches, scientific presentations, thematic workshops, and panel discussions. The conference brought together leading researchers from diverse disciplines to address pressing challenges in IoT and Edge Computing.

We extend our sincere gratitude to all contributors for their dedication and insightful work which made GIECS 2025 a success. Special thanks are due to Damir Filipovic and the entire AIOTI team for their support. The event served as a collaborative platform for exchanging ideas, sharing research, and developing innovative solutions to advance the cloud-IoT-edge continuum.

We look forward to fostering continued dialogue and collaboration in future editions of GIECS.

September 2025

Mirko Presser
Antonio Skarmeta
Srdjan Krco
Dhananjay Singh

Organization

General Chair

Antonio Skarmeta — University of Murcia, Spain

Co-chairs

Mirko Presser — Aarhus University, Denmark
Srdjan Krco — DunavNET, Serbia
Dhananjay Singh — Pennsylvania State University, USA

Program Committee Chairs

Mirko Presser — Aarhus University, Denmark
Aurora Gonzalez Vidal — University of Murcia, Spain

Program Administration

Damir Filipovic — Alliance for AI, IoT and Edge Continuum Innovation (AIOTI), Belgium
Valentina Peniche — Alliance for AI, IoT and Edge Continuum Innovation (AIOTI), Belgium
Emilie Mathilde Jakobsen — Aarhus University, Denmark

Technical Program Committee

Abdellah Zyane — Cadi Ayyad University, Morocco
Asim Ul Haq — Aarhus University, Denmark
David Sarabia — i2CAT Foundation, Spain
Davide Moroni — Institute of Information Science and Technology, Italian National Research Council, Italy
Enrico Fererra — LINKS Foundation, Autonomous IoT Systems and Robotics Connected Systems and Cybersecurity (CSC), Italy

Harris Niavis	Inlecom Innovation, Greece
Konstantinos Loupos	Inlecom Innovation, Greece
Krzysztof Piotrowski	Leibniz Institute for High Performance Microelectronics, Germany
Michael Lystbæk	Aarhus University, Denmark
Nidhi	Aarhus University, Denmark
Parwinder Singh	Aarhus University, Denmark
Tianyue Chu	Telefónica Innovation Digital, Spain
Valentina Tomat	University of Murcia, Spain
Vasileios Karagiannis	Austrian Institute of Technology, Austria

Contents

Smarter IoT: Energy, Connectivity and Real-World Impact

Cloud-Based Interoperability in Residential Energy Systems 3
    Darren Leniston, David Ryan, Ammar Malik, Jack Jackman,
    and Terence O'Donnell

LoRa and Mioty - A Comparative Study 18
    Joerg Robert and Thomas Lauterbach

NeuroKey: A Lightweight AI Tool Using Passive Keystroke Dynamics
for Parkinson's Disease Detection and Monitoring 39
    Ritu Chauhan, Mehak Jena, and Dhananjay Singh

Building Trust: Privacy, Security and Responsible AI

Enhancing Security and Privacy in Federated Learning for Distributed
Systems: The REMINDER Approach 57
    Francisco J. Cortés-Delgado, Enrique Mármol Campos,
    José L. Hernández-Ramos, Antonio Skarmeta, Shahid Latif,
    Djamel Djenouri, Stephan Krenn, Andrei Puiu, and Anamaria Vizitiu

Towards a Responsible AI Adoption/Adaptation (RAA) Ecosystem:
Vision and Model to Keep Socio-Technological Balance 71
    Parwinder Singh, Asim Ul Haq, and Mirko Presser

Distributed and Trusted Access to Data Spaces' Products Employing
Sovity and Data Fabric 93
    Matilde Julian, Miguel Ángel Esbrí, Ignacio Lacalle, Lucía Cabanillas,
    Rafael Vaño, and Carlos E. Palau

Data Spaces and Digital Infrastructure for the IoT Era

Web Based Monitoring, Orchestration and Simulation 111
    Dave Raggett

EOSC and Data Spaces for Cross-Domain Data Sharing in Europe:
Insights from the TITAN-EOSC Project 125
    Natalia Borgoñós García, María Hernández Padilla, Jose Vivo Pérez,
    and Antonio Fernando Skarmeta Gómez

Multi-Agent Stateless Orchestration for Distributed Data Pipelines
Implementation 141
   *Nicolò Bertozzi, Anna Geraci, Marco Sacchet, Enrico Ferrera,
and Claudio Pastrone*

Sustainable Solutions and Applied IoT Innovation

RAPT: AI–Powered IoT Framework for Real-Time Respiratory Disorders
Monitoring and Prediction 159
   Ritu Chauhan, Aarushi Mishra, and Dhananjay Singh

Digital Product Passport as Digital Carrier for Information of Life Cycle
Assessment: A Feasibility Study of Solvolysis on Composite Recycling
for Wind Turbines Blades 175
   Christina Tsitsiva and Michail J. Beliatis

Bridging ESG and Capability Maturity: A Case-Based Artefact
for Industrial Organisations 188
   Lasse Cenholt and Mirko Presser

An NGSI-LD-Based ICT Tool for Data Visualization and Traceability
in Sustainable Supply Chains and Biological Resource Certification 206
   Romain Magnani and Franck Le Gall

A Fair and Lightweight Consensus Algorithm for IoT 224
   Sokratis Vavilis, Harris Niavis, and Konstantinos Loupos

Author Index 241

Smarter IoT: Energy, Connectivity and Real-World Impact

Cloud-Based Interoperability in Residential Energy Systems

Darren Leniston[1(✉)], David Ryan[1], Ammar Malik[2], Jack Jackman[1], and Terence O'Donnell[2]

[1] Walton Institute, South East Technological University, Waterford, Ireland
darren.leniston@waltoninstitute.ie
[2] School of Electrical and Electronic Engineering, University College Dublin, Dublin, Ireland

Abstract. As distributed energy resources (DERs) such as solar PV, batteries and electric vehicles become increasingly prevalent at the edge, maintaining grid stability requires advanced monitoring and control mechanisms. This paper presents a scalable smart grid gateway architecture that enables interoperability between Modbus-based inverters and IEEE 2030.5 cloud-based control systems. The proposed solution leverages Azure cloud services and gateway devices located at the edge, to support dynamic configuration, telemetry ingestion, remote control and Volt-VAR Curve deployment. A microservice architecture ensures flexibility and scalability across diverse deployment scenarios, including both gateway-mediated and direct-to-cloud device communication. Results demonstrate the successful mapping of a Fronius Primo inverter's Modbus registers to IEEE 2030.5-compliant telemetry and control functions. Additionally, we evaluate real-time VVC updates and their impact on local voltage regulation, showcasing dynamic cloud-to-edge control with minimal latency. This work highlights the potential of virtualised, standards-based control infrastructures to support DER integration and active grid participation, while remaining adaptable to evolving smart grid architectures.

Keywords: Smart Grid · Distributed Energy Resources · Edge Computing · IEEE 2030.5 · SunSpec Modbus · IoT Gateway

1 Introduction

As the number of renewable energy sources (RES) increases in tandem with the push from both National and European policymakers to mitigate the cumulative impacts of climate change, the energy sector is undergoing significant transformations to align with initiatives like the European Green Deal. The research described in this paper is designed to support this transition by developing a platform to enable Smart Grid (SG) control architectures. This platform provides an example of how Distribution System Operators (DSO) and other stakeholders in the energy sector can effectively integrate SG devices for compliance with data and communication standards. To further promote the standardisation of SG devices, this research has developed an SG gateway device. This device facilitates communications with SG components such as smart inverters while managing telemetry data and control command execution. Built in accordance with IEEE 2030.5 and Irish DSO network standards and protocols, it will enhance both the system and asset owners' visibility and control over their resources, ensuring interoperability between SG devices [12]. Additionally, the research has implemented a cloud platform architecture to support bi-directional communication with gateway devices in the field. This cloud platform enables intelligent deployment, management, and monitoring of gateway devices in compliance with the requirements of the IEEE 2030.5 standard in terms of interfaces and cybersecurity.

2 Background

2.1 Remote Configuration in the Smart Grid

As the structure of the energy grid becomes increasingly distributed, utilities are implementing and leveraging the capabilities of edge computing hardware to enable monitoring and fine-grained control functionality. In this context, the ability to remotely configure these edge devices is crucial, ensuring the necessary level of flexibility, resilience, and efficiency required in modern distribution systems. Traditionally, applying configuration and firmware updates to smart meters, RTUs (Remote Terminal Units), and other assets located at the network edge required manual intervention, often involving physical access to the device itself [3]. To address these challenges, modern SG systems rely on automated OTA (over-the-air) configuration techniques that leverage cloud computing capabilities, software-defined networking (SDN), and more recently on AI and ML-based technologies. Furthermore, zero-touch-provisioning (ZTP), a paradigm leveraged and described in this paper, has further streamlined remote configuration by enabling edge devices to automatically retrieve and apply configuration settings upon deployment [5]. This eliminates the need for manual setup and reduces operational costs, particularly in large-scale grid roll outs. Cloud-based device management platforms play a vital role in orchestrating these processes, allowing utilities to remotely monitor, configure, and troubleshoot devices in real-time [5]. AI and machine learning techniques can be utilised to further enhance these capabilities by predicting configuration failures, optimising parameter settings, and identifying anomalies that may indicate security threats [6].

2.2 Smart Grid Standardisation and Interoperability

SG interoperability is a key factor for integrating Distributed Energy Resources (DERs) into modern energy grids. An assortment of standards facilitate seamless communication among diverse systems, enhancing grid management, energy distribution and reliability. These standards support the integration of RES, enable demand response capabilities and support Advanced Metering Infrastructure (AMI). Communication standards are particularly critical, enabling the remote configuration and device management across heterogenous infrastructures. For example, IEC 61850 supports substation automation and real-time data exchange, while OpenADR enables dynamic demand response by adjusting devices based on grid conditions [4]. The IEEE 2030.5 standard, part of the Smart Grid Interoperability Series, governs utility interaction with consumer devices via internet protocols such as TCP/IP and a RESTful architecture leveraging XML. It is comprised of "Function Sets" covering DERs, Electric Vehicle (EV) support, billing and demand response. The standard utilises elements from IEC 61968 and supports multiple deployment models, including in-home and cloud-based setups [13]. The Common Information Model (CIM) provides a standardised framework for representing power systems in both UML and XML [14]. The standard includes IEC 61970 (energy management), IEC 61968 (utility domain integration) and IEC 62325 (market communication) with the aim of promoting interoperability and efficiency [14]. SAREF (Appliances) is a standard which provides an ontological view for aligning IoT assets across domains. Its modular structure allows for domain-specific extensions [15], such as SAREF4ENER which models DER flexibility for demand response, in addition to SAREF4GRID which standardises SG IoT data.

2.3 Digital Twins in the Smart Grid

A digital twin serves as a virtual representation of a physical asset or system that updates in real-time using data collected from the edge [8]. In smart grids, digital twins enhance monitoring, simulation and optimisation across power generation, transmission, distribution and consumption contexts, thereby improving efficiency, resilience and decision-making capabilities [9]. A key benefit is predictive maintenance, where real-time data from assets such as transformers enables early fault detection, reducing downtime and associated costs. Digital twins also support contingency analysis, aiding operators in mitigating the impacts from extreme weather or load fluctuations [10]. Their implementation depends on IoT, machine learning and cloud/edge computing, where IoT sensors supply granular asset data, cloud platforms handle large-scale simulations and edge computing ensures low-latency response. Interoperability with physical systems is supported by standards such as CIM, IEEE 2030.5 and IEC 61850 [10]. Extending beyond operational optimisation, digital twins also aid in real-time control, energy market simulations, DER integration and demand response testing. They also bolster cybersecurity by simulating attack scenarios. However, challenges such as high costs, data privacy and governance hinder broader adoption and research is ongoing to mitigate theses concerns by explore solutions such as blockchain-based secure data sharing [11].

3 Use Cases

3.1 Mapping of Inverter Communications to Smart Grid Standards

As protocols such as IEEE 2030.5 and SunSpec Modbus gain traction among DSOs, enhancing edge interoperability has become increasingly vital. The proliferation of residential solar PV systems for example, coupled with the heterogeneity of inverter technologies, both compliant and non-compliant with SunSpec, challenge the DSOs' ability to maintain grid observability and stability. Consequently, new use cases are emerging to standardise telemetry and control interfaces across diverse devices. The use case described here addresses the mapping of Modbus registers to IEEE 2030.5 function sets for edge-to-cloud communication. Modbus, a widely used industrial protocol, follows a request/response model, typically over RS485 or TCP/IP. It leverages four key data types, discrete inputs, coils, input registers and holding registers, to enable access to sensor data and control commands. A suitable smart inverter must support Modbus over TCP/IP and RS485 in addition to offering a set of registers with defined address spaces, data types and control functions. However, register configurations can vary significantly by vendor, complicating interoperability and potentially causing communication failures with upstream platforms when data formats or mappings are inconsistent. This use case demonstrates how aligning Modbus register functions with IEEE 2030.5 function sets can establish a unified data layer between inverter telemetry/control systems and cloud-based grid management platforms, enabling more reliable integration and household participation in DSO-led grid balancing efforts. Figure 1 illustrates this mapping, where functions are exposed through an IEEE 2030.5 API, linking a PV deployment to an IEEE 2030.5 compliant system while utilising Modbus registers to interface with inverter functions.

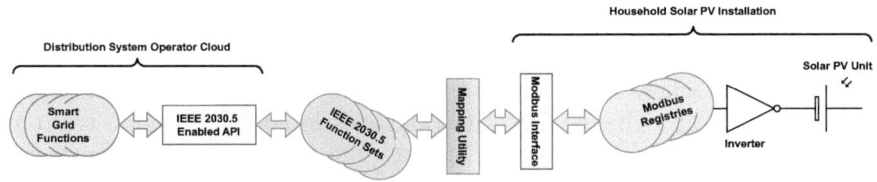

Fig. 1. Solar PV to Smart Grid Communications mapping

3.2 Semi Autonomous DER Control

As smart grids evolve, more consumers begin to act as prosumers, contributing grid services via DERs such as solar PV, batteries and Vehicle-to-Grid (V2G) enabled EVs. To prevent instability from widespread DER adoption, for example voltage fluctuations, frequency imbalances and reverse power flows, smart inverter-level control is essential. Modern inverters support Volt/VAR and Volt/Watt control, enabling voltage regulation in real-time. These capabilities, combined with bi-directional communication via Advanced Distribution Management Systems (ADMS) and Distributed Energy Management Systems (DERMS), allow dynamic adjustments that enhance grid resilience. Unlike simple curtailment, Volt-VAR Curves (VVCs) dynamically adjust reactive power in response to grid conditions. VVCs are particularly valuable in high DER or EV penetration areas, mitigating voltage fluctuations caused by intermittent generation and bi-directional energy flow, ensuring stability without reducing active power exports. The use case described here explores the cloud-based generation of VVCs for clusters of DERs, and their delegation to inverter-level control. VVCs are scheduled with grid demand and capacity in mind, requiring systems that can distribute updates to edge devices. While some inverters can directly implement curves, others rely on receiving reactive power setpoints, introducing complexity in scaling across diverse devices. Figure 2 illustrates generation of the VVC in the DSO's cloud environment and transmission to a household DER installation for actuation, whether that be the inference of a reactive power setpoint, as in this case, or the autonomous programming of the VVC on the inverter.

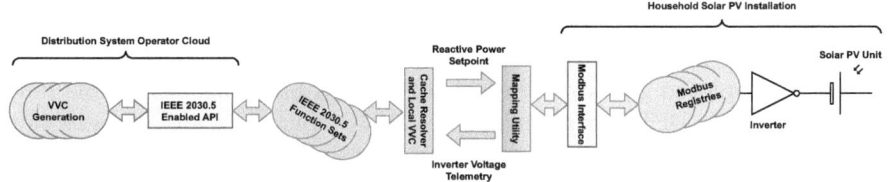

Fig. 2. Cloud to Inverter VVC Transfer

4 Proposed Remote Configuration System for Interoperability and Standards Management

4.1 Edge Device Design

The concept for an SG gateway is based on the IoT gateway model, acting as a central hub for communication with devices such as PV inverters and smart meters. These gateways and their functionality are critical across both distribution and transmission level, facilitating large-scale data retrieval and integration of diverse platforms across Low Voltage (LV), Medium Voltage (MV) and High Voltage (HV) layers. Depending on their capabilities, gateways also support edge computing, enabling local data processing and aggregation to reduce network traffic and latency. Additionally, their customisable software and firmware, coupled with the capability of remotely updating the device supports multiple use cases. Unlike more general IoT devices, SG devices require strict compliance with communication standards such as IEEE 2030.5 and (SunSpec) Modbus, which define common parameters for DER monitoring and control [12]. Thus, any gateway must support SunSpec modbus at minimum, in addition with other necessary protocols. Hardware requirements for such a gateway include RS485 ports for serial connections, wireless/Ethernet for internet-based comms and/or cellular modules for remote connectivity. Gateways must be physically secure to protect data and grid assets and robust enough for deployment in the field.

Typically running Unix-based systems, they support containerised microservices for flexible deployment and often integrate natively with cloud platforms or offer tailored application support.

4.2 Edge Gateway Design and Flow

At the core of the SG Gateway concept is an edge-level microservice architecture that enables flexible configuration and control delegation. Unlike monolithic systems, microservices are loosely coupled and independently deployable, and as such enhance robustness, scalability and maintainability. Each service encapsulates specific logic for telemetry, remote configuration and device control, supporting modular edge functionality. Gateway devices are orchestrated via a cloud-based SG management platform, which leverages cloud scalability to handle telemetry data, dispatch control signals, perform software updates and manage device mappings to IEEE 2030.5 and Volt-VAR curves. This architecture offloads complex logic from the edge to the cloud, simplifying the management of numerous deployed gateways. Key cloud-enabled functions include SG device telemetry capture, DER curtailment, Local VVC operation, and updates to the gateway device cache for IEEE 2030.5 mappings and VVC bounds. Communication is facilitated using Azure Direct Methods, allowing device-specific actions to be triggered through an IoT Hub via REST compliant requests [16]. Figure 3 illustrates how device functions are invoked on a gateway device.

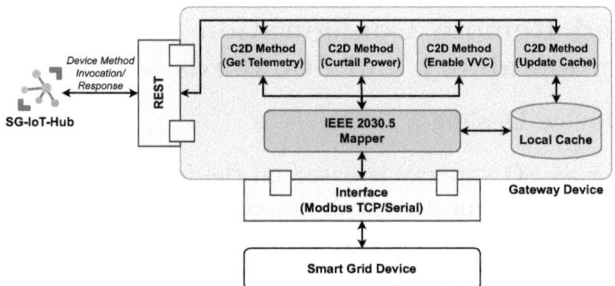

Fig. 3. Device method invocation flow

In order for a request to be valid, it must be authenticated through the cloud platform before being invoked on a device or suite of devices. For example, a request for telemetry on a specific register through a gateway device would be made via POST request.

Curtail Power: The API endpoint defined for this action on the gateway is determined by specifying the device to be triggered (e.g., "gateway_device_XYZ") invoking the method "curtailPower" and passing a write command with a Modbus register and value, with a request made to this endpoint triggering the data flow presented in Fig. 4. This command curtails active power on the connected SG device (e.g., inverter) by a given percentage. These control signals are defined in the cloud system to be actuated on the connected SG device through the gateway.

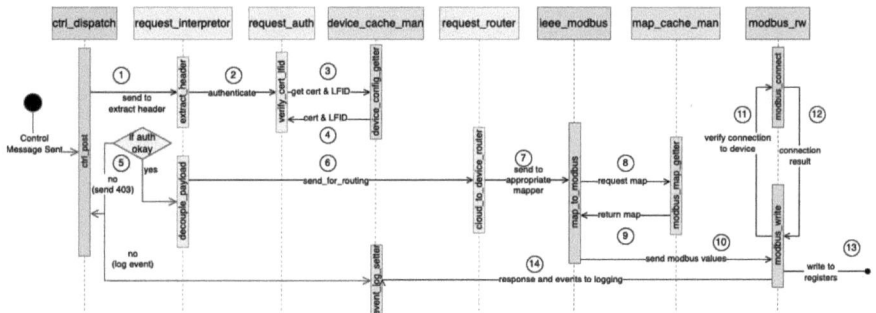

Fig. 4. Cloud to device control signal data flow

Update Gateway Cache: Requests to this endpoint initiate the dataflow illustrated in Fig. 5. Mapping specifications transmitted from the cloud define how data from specific SG devices is aligned with IEEE 2030.5 for purposes including control decision-making, monitoring and historical data storage. To support dynamic updates, a local cache is maintained on each gateway device, offering high-performance, small-scale storage. Cache entries are structured as dictionaries and device methods enable management of this cache. For example, the "updateGatewayCache" function updates mappings by specifying a device type and associated mapping. Gateways are initially provisioned with a base mapping version, which can be seamlessly updated via the cloud to accommodate changes in IEEE 2030.5, thereby enhancing system adaptability and future-proofing. This mechanism also enables dynamic updates to the local VVC, allowing refined control directly to the edge.

Device to Cloud Telemetry: This dataflow governs device-to-cloud communication for capturing SG device telemetry and is initiated via the "getTelemetry" device method. Upon invocation, the gateway device retrieves specified values from the Modbus telemetry registers, maps them to the corresponding IEEE 2030.5 entries, and transmits the data through a central message hub to the SG cloud platform for processing and storage. Figure 6 illustrates the sequence triggered by any valid telemetry value.

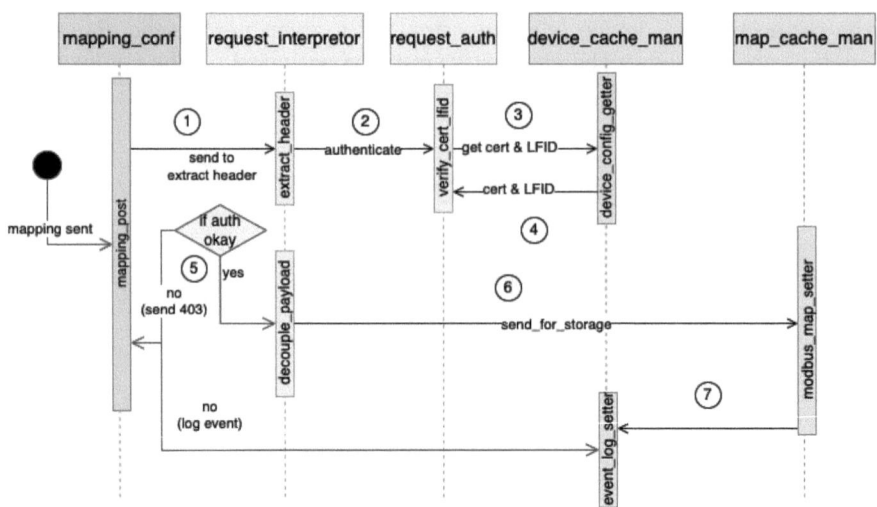

Fig. 5. Cloud to device mapping data flow

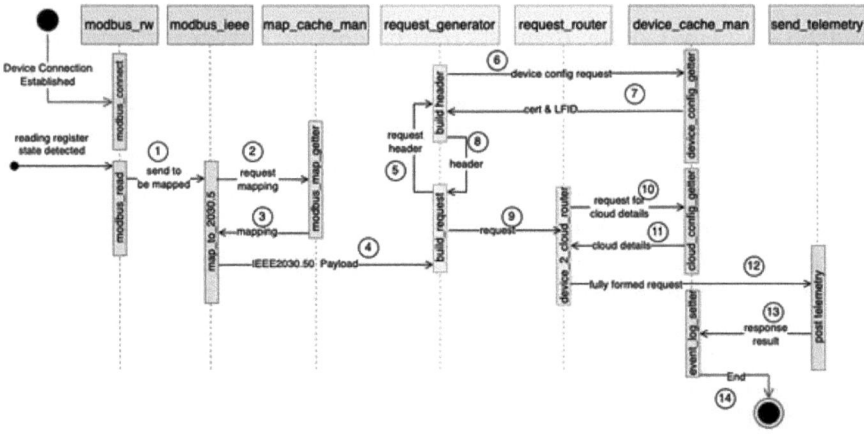

Fig. 6. Device to cloud telemetry data flow

4.3 Cloud Twin Implementation

Within this section we outline the cloud architecture, built on Microsoft Azure's Infrastructure as a Service (IaaS) platform, to support gateway device management, data acquisition, control, processing, storage and visualisation. The system includes APIs for telemetry retrieval, control signal transmission and SG device mapping management, as illustrated in Fig. 7.

IoT Hub: Serves as the central access point, handling authentication, bidirectional data flow, device twin updates and IEEE 2030.5-compliant telemetry ingestion. **Device-Telemetry-Ingress**: Is an Azure Function triggered

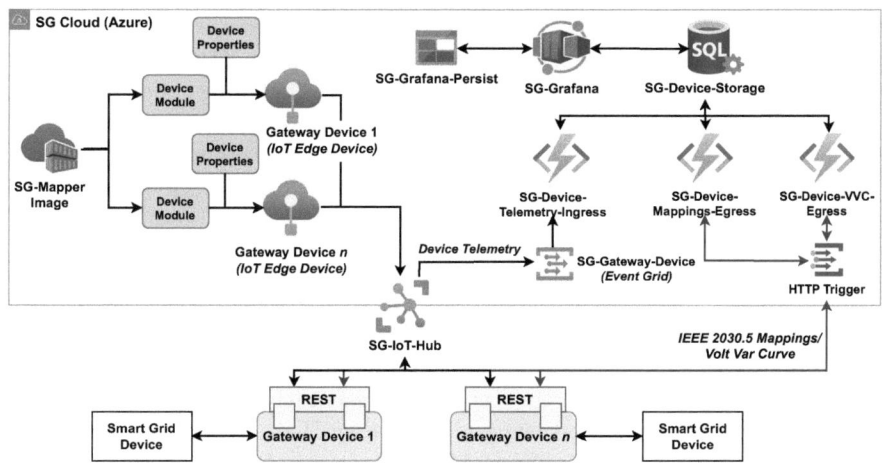

Fig. 7. Cloud system architecture

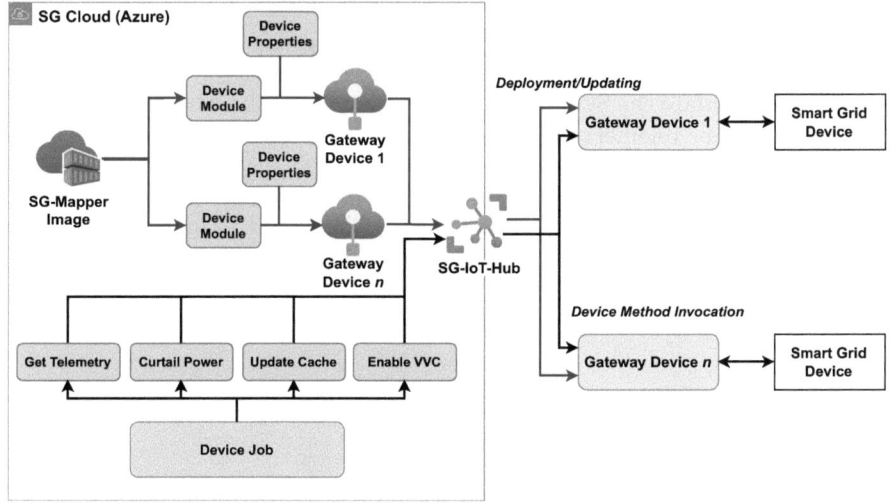

Fig. 8. Gateway device deployment and scalability architecture

by the Gateway-Device Event Grid to parse and store telemetry. **Device-Mappings-Egress**: Responds to authenticated REST requests, delivering IEEE 2030.5-compliant mappings. **Device-VVC-Egress**: Responds to authenticated REST requests, returning VVC bounds. **Device-Storage**: Is an SQL-based repository for telemetry, gateway metadata, standard mappings and VVC bounds. **Dashboard-Service**: Leverages Grafana (deployed via Azure Container Apps) to visualise telemetry from Device-Storage. **Dashboard-Service-Persist**: Retains Grafana settings such as dashboards and user accounts using an external SQLite database. **Gateway-Registry**: Hosts container images for gateway device software across hardware variants via Docker. **IoT Edge**: Provisions and monitors deployed edge devices, each with a Device Twin for configuration. The Device Module defines the deployed software, such as mapping containers, while Device Properties specify operational settings such as REST endpoint URLs, connected device types, IP addresses and communication modes. IoT Edge interaction with gateway devices in the field is described in Fig. 8.

5 Results

To support testing of the SG gateway device and cloud platform, a Hardware-in-the-Loop (HIL) laboratory environment was utilised consisting of a Fronius Primo single-phase inverter acting as the representative SG device. The smart grid gateway was implemented using a Raspberry Pi 4, configured to communicate with the inverter over SunSpec Modbus via wireless TCP/IP connection. This setup served as the basis for evaluating both mapping of telemetry and the dynamic control functionalities of the system.

5.1 Mapping Smart Inverter Telemetry to IEEE 2030.5

In an effort to validate the SG platform's mapping layer, the laboratory setup was leveraged to gather telemetry data from the inverter (SG device) over TCP/IP through the gateway. This data included data points such as AC active power, AC voltage and reactive power, retrieved in real-time using SunSpec compliant Modbus registers. These data points were then mapped to IEEE 2030.5 compliant telemetry structures using the mapping service hosted on the gateway. A typical example is presented in Table 1, where telemetry values from the Modbus interface are mapped to their respective IEEE 2030.5 fields.

Table 1. Example Mapping from SunSpec Modbus Registers to IEEE 2030.5 Telemetry Fields

Parameter	SunSpec Modbus Register	IEEE 2030.5 Field Path	Units
Active Power (W)	40083	DERStatus/W	Watts
AC Voltage	40072	DERStatus/V	Volts
AC Current	40076	DERCapability/Amp	Amperes
Frequency	40070	DERStatus/Hz	Hertz
Reactive Power	40084	DERStatus/VAR	VAR

A representative JSON payload produced by the gateway for cloud ingestion is presented in listing 1.1, which describes an IEEE 2030.5 mapping for an Active Power telemetry reading gathered from the inverter in real-time.

Listing 1.1. Example IEEE 2030.5 Reactive Power Reading

```
[
  {
    "40083": {
      "ReadingType": {
        "description": "W (Active power)",
        "accumulationBehaviour": 12,
        "commodity": 1,
        "dataQualifier": 0,
        "flowDirection": 1,
        "intervalLength": 3600,
        "kind": 0,
        "numberOfConsumptionBlocks": 0,
        "numberOfTouTiers": 0,
        "phase": 0,
        "powerOfTenMultiplier": 0,
        "tieredConsumptionBlocks": "false",
        "uom": 38
      },
      "Reading": {
        "consumptionBlock": "0 - N/A",
        "qualityFlags": "01",
        "timePeriod": {
          "duration": 60,
          "start": 1767148190
        },
        "touTier": "0 - N/A",
        "value": 1914
      }
    }
  }
]
```

Figure 9 presents a sample of the SG Cloud Platform visualisation of gathered telemetry readings from the HIL system, including device location, AC Power, Voltage, AC Energy and Line Frequency.

5.2 Dynamic Volt-VAR Curve Update and Execution

A key element of the work described in this paper was validating the ability to dynamically deploy and execute Volt-VAR Curves during real-time grid simulation. This test utilised the laboratory HIL system to emulate changes in voltage conditions in a controlled manner, with real-time curve logic handled by the gateway device. Two separate Volt-VAR Curves were defined for the experiment to limit VAR injection, these curves were ingested by the gateway device in JSON format. Listing 1.2 describes the operational bounds of the first Volt-VAR Curve (VVC1) tested, while Listing 1.3 describes the operational bounds of the second curve (VVC2) tested.

Listing 1.2. Volt-VAR Curve 1 Bounds

```
{"VOLTAGE_THRESHOLDS": {"V90": 207.0, "V95": 218.5, "V105": 241.5, "V110": 253.0}, "REACTIVE_POWER_LIMITS": {"Q_EXPORT": 45.4, "Q_ABSORB": -45.4, "UNITY": 0.0}, "ACTIVE_POWER_LIMIT": 75.0}
```

Listing 1.3. Volt-VAR Curve 2 Bounds

```
{"VOLTAGE_THRESHOLDS":{"V90":207.0,"V95":218.5,"V105":241.5,"V110":253.0},"REACTIVE_POWER_LIMITS":{"Q_EXPORT":40.5,"Q_ABSORB":-40.5,"UNITY":0.0},"ACTIVE_POWER_LIMIT":100.0}
```

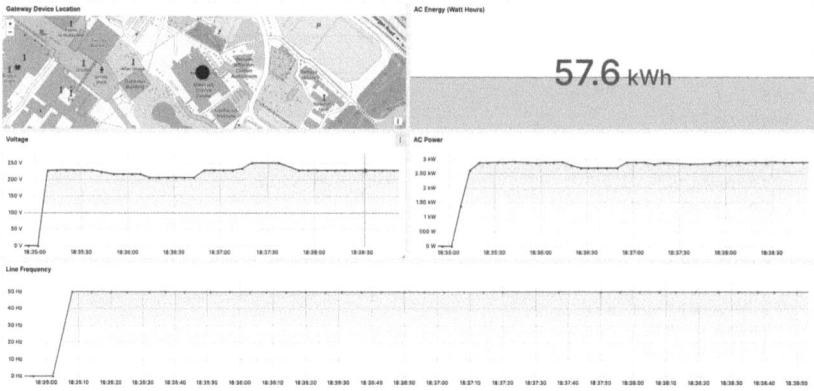

Fig. 9. SG Platform visualisation dashboard

Initially, VVC1 was active on the gateway, midway through the grid simulation a new curve (VVC2) was dynamically pushed to the gateway device through the "updateGatewayCache" method described in Sect. 4.2 (Edge Gateway Design and Flow). The curve bounds were sent from the cloud to the gateway, which parsed the JSON structure to generate reactive power setpoints that were then applied to the inverter via Modbus TCP/IP. Figure 10 presents a time-series plot of voltage and reactive power over the testing period. The moment of the VVC update is indicated, after which a clear shift in inverter behaviour can be observed through the peaks and troughs in the Q (Var), which are reduced under VVC2 post-update, reflecting the intended adjustment in reactive power output.

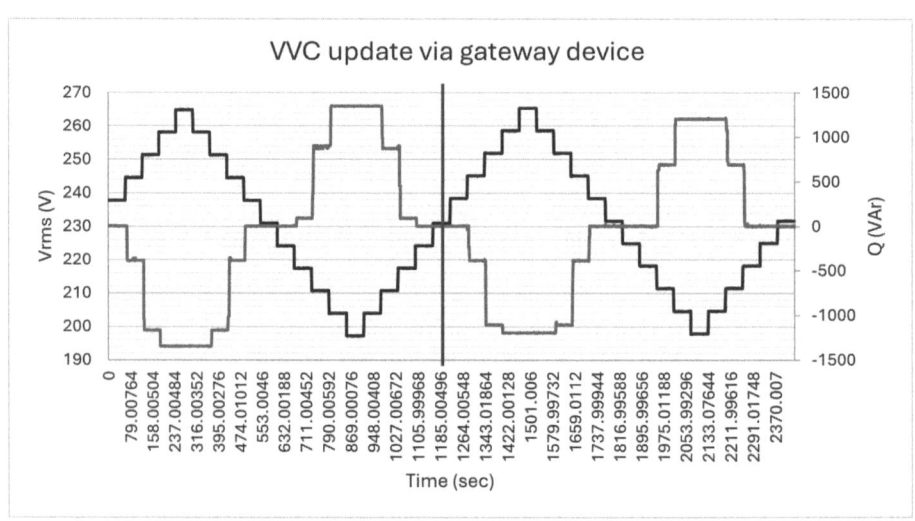

Fig. 10. Reactive power response before and after VVC update

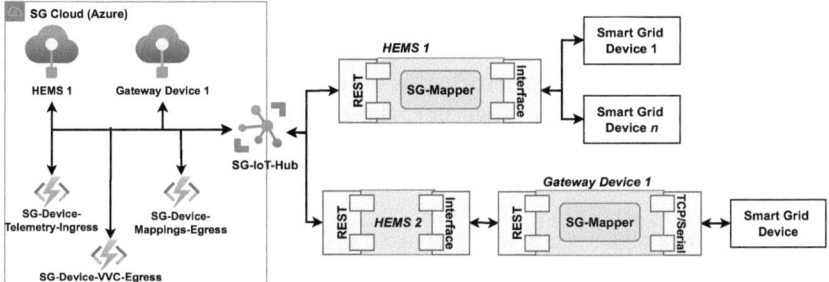

Fig. 11. Cloud architecture to support HEMS/third-party cloud systems

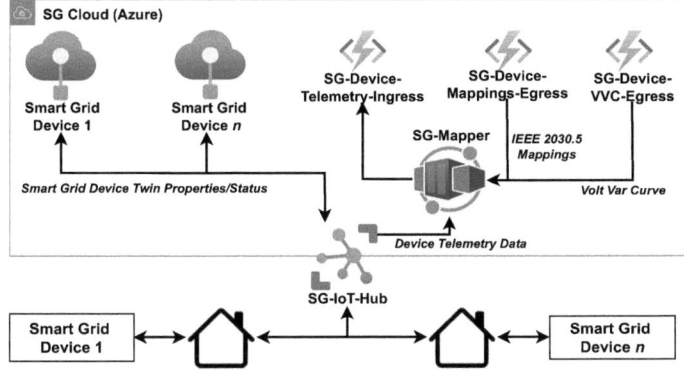

Fig. 12. Cloud architecture to support native smart grid device control

This confirms that the update was successfully applied and executed. This test demonstrates the systems ability to dynamically modify grid-support behaviours in real-time and seamless communication from cloud to edge to device.

6 Conclusion and Future Work

This work has demonstrated a scalable smart grid gateway and cloud platform that bridges Modbus-based inverters with IEEE 2030.5-compliant control systems. Using a microservice architecture with Azure integration and edge gateways, the solution supports dynamic configuration, telemetry ingestion, and remote control of DERs. Hardware-in-the-Loop testing with a Fronius Primo inverter validated the mapping of Modbus telemetry to IEEE 2030.5 structures (Table 1) and confirmed successful deployment of dynamic Volt-VAR Curve updates with minimal latency (Fig. 10), highlighting the potential of standards-based, virtualised infrastructures for interoperable and responsive grid support. Beyond the gateway-mediated model, alternative architectures are emerging. As shown in Fig. 11, the gateway mapper could potentially run directly within a HEMS managing multiple devices, while Fig. 12 illustrates a cloud-centric model where inverters connect natively, eliminating the locally deployed gateway. Thanks to its modular, containerised design, the platform can adapt to these scenarios with limited refactoring, requiring only enhancements for multi-device support and scalability. Together, the results and design flexibility point to a pathway for interoperable, secure, and future-ready smart grid control architectures that can accommodate the growing integration of DERs.

Acknowledgments. This study was supported by SEAI (Sustainable Energy Authority of Ireland) in collaboration with ESB Networks, with contributions from SIRFN and OPENSVP. Additional support was provided through the COALESCE project under the European Union's Horizon Europe programme, Marie Skłodowska-Curie Actions.

References

1. Author, A.-B.: Contribution title. In: 9th International Proceedings on Proceedings, pp. 1–2. Publisher, Location (2010)
2. LNCS Homepage. http://www.springer.com/lncs. Accessed 25 Oct 2023
3. Lekbich, A., Belfqih, A., Zedak, C., Boukherouaa, J., El Mariami, F.: A secure wireless control of remote terminal unit using the internet of things in smart grids. In: 2018 6th International Conference on Wireless Networks and Mobile Communications (WINCOM), Marrakesh, Morocco, pp. 1–6 (2018). https://doi.org/10.1109/WINCOM.2018.8629620
4. Ninagawa, C., Iwahara, T., Suzuki, K.: Enhancement of OpenADR communication for flexible fast ADR aggregation using TRAP mechanism of IEEE1888 protocol. In: 2015 IEEE International Conference on Industrial Technology (ICIT), Seville, Spain, pp. 2450–2454 (2015). https://doi.org/10.1109/ICIT.2015.7125458
5. Demchenko, Y., et al.: ZeroTouch Provisioning (ZTP) model and infrastructure components for multi-provider cloud services provisioning. In: 2016 IEEE International Conference on Cloud Engineering Workshop (IC2EW), Berlin, Germany, pp. 184–189 (2016). https://doi.org/10.1109/IC2EW.2016.50
6. Hazra, A., et al.: Distributed AI in zero-touch provisioning for edge networks: challenges and research directions. Computer **57**(3), 69–78 (2024). https://doi.org/10.1109/MC.2023.3334913
7. Feng, G., Huang, Q., Deng, Z., Zou, H., Zhang, J.: Research on cloud security construction of power grid in smart era. In: 2022 IEEE 2nd International Conference on Data Science and Computer Application (ICDSCA), Dalian, China, pp. 976–980 (2022). https://doi.org/10.1109/ICDSCA56264.2022.9987863
8. Baek, M.-S.: Digital twin federation and data validation method. In: 2022 27th Asia Pacific Conference on Communications (APCC), Jeju Island, Korea, Republic of 2022, pp. 445–446 (2022). https://doi.org/10.1109/APCC55198.2022.9943622
9. Lu, Q., Jiang, H., Chen, S., Gu, Y., Gao, T., Zhang, J.: Applications of digital twin system in a smart city system with multi-energy. In: 2021 IEEE 1st International Conference on Digital Twins and Parallel Intelligence (DTPI), Beijing, China, pp. 58–61 (2021). https://doi.org/10.1109/DTPI52967.2021.9540135
10. Balijepalli, V.M., Sielker, F., Karmakar, G.: Evolution of power system CIM to digital twins - a comprehensive review and analysis. In: 2021 IEEE PES Innovative Smart Grid Technologies Europe (ISGT Europe). Espoo, Finland 2021, pp. 1–6 (2021). https://doi.org/10.1109/ISGTEurope52324.2021.9640174

11. Koeva, D., Kutkarska, R.: Application of AI to develop a digital twin of energy communities: conceptual approach. In: 2024 5th International Conference on Communications, Information, Electronic and Energy Systems (CIEES), Veliko Tarnovo, Bulgaria, pp. 1–4 (2024). https://doi.org/10.1109/CIEES62939.2024.10811186
12. Johnson, J., Fox, B., Kaur, K., Anandan, J.: Evaluation of interoperable distributed energy resources to IEEE 1547.1 using SunSpec Modbus, IEEE 1815, and IEEE 2030.5. IEEE Access **9**, 142129–142146 (2021). https://doi.org/10.1109/ACCESS.2021.3120304
13. IEEE standards association: IEEE standard for smart energy profile application protocol. IEEE Std. 2030.5-2018 (2018). https://doi.org/10.1109/IEEESTD.2018.8403867
14. Crimmins, S.: Common Information Model CIM Primer, (8th ed.), Electric Power Research Institute (EPRI) (2022)
15. Haghgoo, M., Daniele, L., Nikiforakis, A., de Vos, M., Wissing, C.: SARGON - smart energy domain ontology. IET Smart Cities **2**(4), 191–198 (2020). https://doi.org/10.1049/iet-smc.2020.0049
16. Shi, J., Jin, L., Li, J.: The integration of azure sphere and azure cloud services for internet of things. Appl. Sci. **9**(13), 2746 (2019). https://doi.org/10.3390/app9132746
17. Government of South Australia: Technical regulator guideline: remote communications capabilities for inverters—declared components of electricity infrastructure or an electrical installation associated with an electricity generating plant; deemed methodologies for remote disconnection and reconnection of electricity generating plants; remote updating methods; export limiting methods (2023)

Open Access This chapter is licensed under the terms of the Creative Commons Attribution 4.0 International License (http://creativecommons.org/licenses/by/4.0/), which permits use, sharing, adaptation, distribution and reproduction in any medium or format, as long as you give appropriate credit to the original author(s) and the source, provide a link to the Creative Commons license and indicate if changes were made.

The images or other third party material in this chapter are included in the chapter's Creative Commons license, unless indicated otherwise in a credit line to the material. If material is not included in the chapter's Creative Commons license and your intended use is not permitted by statutory regulation or exceeds the permitted use, you will need to obtain permission directly from the copyright holder.

LoRa and Mioty - A Comparative Study

Joerg Robert[1]( ) and Thomas Lauterbach[2]( )

[1] Friedrich-Alexander-Universität Erlangen-Nürnberg, Erlangen, Germany
`joerg.robert@fau.de`
[2] Technische Hochschule Nürnberg Georg Simon Ohm, Nuremberg, Germany
`thomas.lauterbach@th-nuernberg.de`

Abstract. This paper presents a comprehensive comparison between two leading Low Power Wide Area Network (LPWAN) technologies: LoRa and mioty. While both are designed for energy-efficient, long-range communication in IoT applications, they differ significantly in their modulation schemes, spectral efficiency, robustness, and network capacity. The study combines theoretical analysis with extensive laboratory measurements to evaluate critical performance metrics such as on-air time, packet error rate (PER), sensitivity, and achievable network throughput. LoRa's performance is analyzed across various spreading factors, highlighting its limitations under high-load conditions in dense deployments. Energy consumption per packet and estimated device lifetime under typical usage scenarios are also assessed. The results show that mioty operates closer to theoretical performance limits and offers substantial advantages for large-scale IoT deployments. These findings provide practical insights for network designers selecting between LoRa and mioty, particularly in dense or capacity-critical environments.

Keywords: LoRa · mioty · Low Power Wide Area Networks

1 Introduction

Low Power Wide Area Networks (LPWAN) are a class of networks that offers long-range connectivity with low transmit powers and low cost. LoRa and mioty are two representatives of this class. These networks can be operated like a classical WLAN network without any licensed frequency bands, but with significantly longer range. LoRa is the most widely known LPWAN system. Its modulation scheme is proprietary and not officially disclosed to the public. The specifications above the physical layer are publicly available in the LoRaWAN specification [4]. In contrast, mioty is an open standard specified in ETSI TS 103 357 [5]. It uses a classical Frequency Shift Keying (FSK) modulation that is supported by many sub-1 GHz chipsets available on the market.

LoRa has gained significant attention within the scientific community, and many papers analyze its performance through theory and simulations. In contrast, mioty has not yet achieved comparable visibility. Consequently, only a few scientific publications on the performance of mioty are currently available.

This paper aims to address this gap by comparing the performance of LoRa and mioty based on both theoretical analysis and experimental measurements. While both systems also support the downlink, the scope of this paper is limited to the uplink. Therefore, approaches such as Adaptive Data Rate (ADR), which may be employed in real networks, are not considered, as their analysis would go significantly beyond the scope of a single paper.

The remainder of this paper is structured as follows: Section 2 provides a brief overview of the mioty system. Section 3 examines the on-air time and spectral footprint of both technologies. Section 4 presents the measurement setup and detailed results regarding sensitivity and network capacity. Finally, Sect. 5 concludes the paper.

2 A Brief Overview of Mioty

There are already many scientific documents available that give a comprehensive introduction to the LoRa physical layer, e.g., [16]. Therefore, we will purely focus on mioty in this section.

mioty – standardized as ETSI TS 103 357 TS-UNB [5, Sec. 6] – is an open standard LPWAN published by the European Telecommunications Standards Institute (ETSI) in 2018. mioty is focusing on an asymmetrical star-topology with very cheap and low complexity sensor nodes, and a gateway that is implemented based on a software-defined radio approach. Like for most LPWAN, the focus of the communication is on the uplink from the sensor nodes to the gateway, while a downlink is also supported.

2.1 Mioty Physical Layer

mioty uses Minimum Shift Keying (MSK), a specific configuration of classical frequency shift keying (FSK). It can be generated by practically all state-of-the-art sub-GHz transceiver chips supporting the IEEE 802.15.4 FSK modes [2]. Further, MSK is a linear modulation that allows correlation receivers operating on the theoretical bounds [22]. The symbol rate of the modulation is 2380.371 Sym/s. This rate has been selected as it can be very precisely generated by almost all transceiver chips with commonly used crystal oscillator frequencies. According to the specification the modulator shall use a filtered variant of MSK called Gaussian Minimum Shift Keying (GMSK) with parameter BT = 1.0 to reduce unwanted emissions into adjacent channels. The resulting bandwidth of the MSK signal is in the order of 2.9 kHz [22].

The ultra-low bandwidth of the signal in addition to frequency selective communication channels would significantly reduce the performance [20]. Therefore, uplink packets are divided into at least 24 radio bursts (also called hops), each lasting 15.14 ms and carrying the equivalent of 8 payload bits. Figure 1 shows the principle. The radio bursts are then distributed over a bandwidth of 60 kHz (normal mode) or 720 kHz (wideband mode), offering significant diversity gain. The overall accumulated on-air time is only 360 ms for a typical packet, while

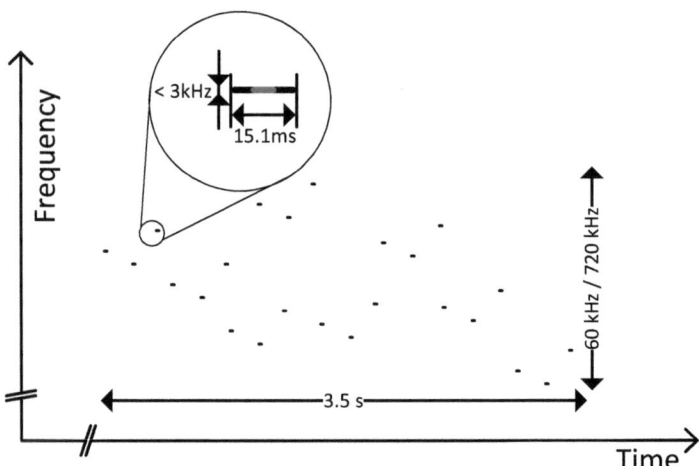

Fig. 1. mioty physical layer uplink: mioty splits the data of one packet into at least 24 radio bursts (hops) that are then sequentially transmitted on different frequencies. The pauses between the radio bursts differ between the normal mode and the low-latency mode. In normal mode, one transmission requires approx. 3.5 s, while the actual transmit duration is approx. 350 ms only. Furthermore, one packet covers a total bandwidth of approx. 60 kHz (EU1 modes) or 720 kHz (EU2 modes).

the pauses between the radio bursts lead to an overall packet duration of 3.5 s in case of the normal modes. The gaps between the radio bursts allow a reduced peak current when the system is operated using tiny coin type batteries [19]. Additionally, the pauses improve frequency and time diversity, and make the system robust against interference from other systems also operated in the same license exempt frequency band [18]. A low-latency mode is also available, which significantly reduces the packet duration by shorter pauses between the radio bursts, resulting in a transmission time in the order of 1 s. However, this comes at the cost of higher peak current and reduced diversity in difficult channels. To be able to recover lost radio bursts mioty employs a forward error correction (FEC) based on a rate 1/3 convolutional code [22] that spreads the encoded data equally over all radio bursts.

Each radio burst carries a midamble of 12 known symbols for synchronization. However, this does not enable the reliable detection of each individual radio burst [29]. Therefore, all radio bursts have to be transmitted in a well-defined time and frequency pattern to enable a joint synchronization within the gateway. The overall timing error over the time-frequency-pattern should be less than 1/4 symbol duration, i.e. 0.1 ms. Otherwise degradation of the MSK decoding can be expected [22]. This results in a required frequency oscillator accuracy of better than 28 ppm (parts per million) for the symbol rate generation, which is supported by low-cost crystal oscillators. Generally, the gateway is based on a software defined radio (SDR) for improved decoding. The main difficulty is in the detection of the signals, which is achieved by digital signal processing in the SDR gateway. Further, the gateway has to be able to decode multiple simultaneously transmitted packets, as the overall network capacity would be very small otherwise.

mioty offers different modes for the transmission. Recommended modes are given in the annex of the ETSI standard [5, Annex B]. The modes only scale the spacing between the bursts, while the modulation is always 2380.371 Sym/s, resulting in a burst bandwidth of less than 3 kHz. Most relevant are the modes EU1, EU2, and US0. Mode EU1 transmits the mioty signals within two frequency bands of 60 kHz (normal mode), while each packet is only transmitted within a single band. The modes EU2 and US0 scale the bandwidth of the frequency bands by a factor of 12 (wideband mode) leading to two frequency bands with 720 kHz each. This offers higher robustness in frequency selective channels and is required to fulfill the frequency regulation in the US. However, the higher bandwidth requires higher computational resources in the gateway compared to the normal mode. The normal as well as the wideband mode can be further combined with the low latency transmission, reducing the pauses between the bursts.

For the downlink mioty supports two different waveforms. One waveform is a classical FSK waveform that can be decoded by the hardware of practically all sub-GHz IEEE 802.15.4 compliant chips. The recommended downlink waveform is very similar to the uplink waveform and splits the data into several bursts. This also assumes a SDR receiver inside the sensornode, resulting in a gain in the order of 9 dB compared to the classical waveform [5, Table 6-1].

Using SDR decoding inside the sensor node seems quite complex. However, the low bandwidth of the modulation in addition to the well-known time and frequency positions highly reduces the amount of data that have to be processed. Therefore, the SDR decoding is possible with state-of-the-art low-cost sub-GHz transceivers, e.g. from Texas Instruments[1].

2.2 Mioty MAC Layer

mioty supports two MAC types, i.e. the fixed MAC mode and the variable MAC. The fixed MAC is fully defined within the standard [5, Sec. 6.3.2]. It offers all required means for the transmission of payload data, signaling information, authentication and encryption. Similar to LoRaWAN, the addressing scheme uses 8 bytes (64 bit) addresses based on the EUI-64 addresses administrated by IEEE. These EUI-64 addresses contain a unique manufacturer and device identification. In order to reduce the address overhead from 8 bytes to 2 bytes a transparent scheme called "short addressing mode" can be used. The encryption and authentication are based on the Advanced Encryption Standard (AES) with 128 bits [1], which is also used by LoRaWAN. This block cipher offers state-of-the-art security with low computational complexity. Further, many IEEE 802.15.4

---

[1] https://www.ti.com/tool/MIOTY, accessed: June 2025.

compliant chips also support hardware acceleration. The typical overhead of the fixed MAC for an uplink payload transmission is 10 bytes.

The variable MAC [5, Sec. 6.3.3] offers support for MAC types defined outside the original ETSI standard. It would be possible to run the LoRaWAN MAC over the mioty physical layer.

3 On-Air Time and Spectral Footprint

For the comparison of the performance of the two systems the on-air time as well as the spectral footprint are of high relevance. The on-air time is the actual time the transmitter is in transmit mode. For a given transmit power, it hence directly affects the required energy to transmit one packet, directly affecting the battery lifetime.

The spectral footprint η is especially relevant for the co-existence of multiple networks. It is the product of the on-air time and the effective system bandwidth, and shows the amount of spectral resources used by the system. Minimizing this value increases the overall network capacity in a given geographic area. The detailed derivations of the on-air time as well as the spectral footprint η are given in the appendix.

The symbol rate of mioty is fixed to 2380.371 Sym/s in all modes. Consequently, the only parameter that effects the on-air time is the data length. Payload data up to 13 payload bytes only use the core frame consisting of 24 radio bursts. This results in a realistic on-air time of 403 ms for packets of up to 13 payload bytes. If more than 13 payload bytes are transmitted, the on-air-time increases by 16.8 ms for every additional byte. The spectral footprint of a mioty packet with up to 13 payload bytes is given by $\eta_{\text{mioty,real}} = 1.2\,\text{kHz} \times \text{s}$.

The on-air time of LoRa depends on a list of different parameters. Details are again given in the appendix. For the following calculations we use the typical European parameter set defined by the "The Things Network" (TTN) – a popular community LoRaWAN network [11].

The on-air time differs significantly between the different SFs. Assuming a payload of 10 bytes, SF7 requires an on-air time of only 62 ms, which increases to 1483 ms for the most robust mode, SF12. The spectral footprint for SF7 is $\eta_{\text{LoRa,SF7}} = 7.7\,\text{kHz} \times \text{s}$, which means that SF7 requires approximately 6.5 times more spectral resources than mioty. For SF12, the spectral footprint increases to $\eta_{\text{LoRa,SF12}} = 185\,\text{kHz} \times \text{s}$, meaning that LoRa SF12 requires about 150 times more of the scarce spectral resources compared to mioty.

Figure 2 shows the on-air times of mioty and LoRa for different parameter configurations and payload lengths. The on-air time of LoRa increases with the spreading factor and the payload length. The minimum realistic on-air time for SF12 is approx. 1.4 s. Generally, the on-air time shows an almost linear increase in addition to a fixed offset. The block interleaver used by LoRa results in the step-like pattern. Especially for the high spreading factors very long on-air times can be reached. However, this may violate the frequency regulation. Some frequency bands in Europe have a maximum duty cycle of 0.1% [6, Annex B]. Hence,

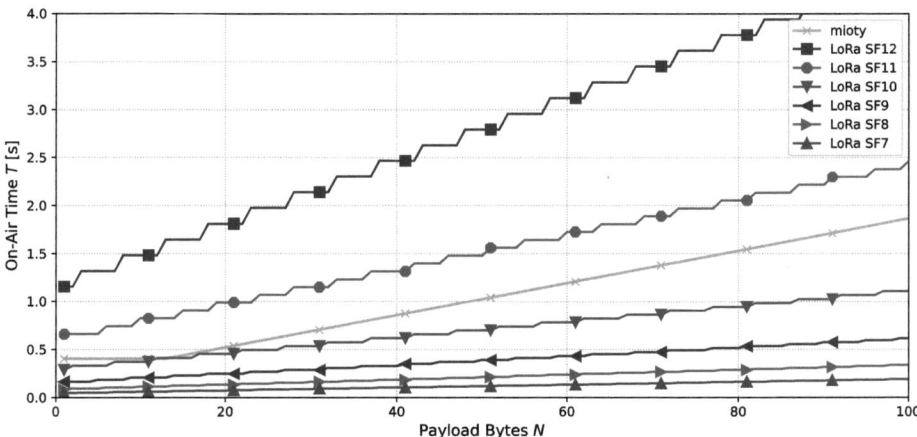

Fig. 2. On-air times of mioty and LoRa (European TTN parameter set) as a function of the number of payload bytes. The staircase structure for LoRa is caused by the block interleaver that adds padding to achieve a specific data length.

the maximum transmit duration per hour is limited to 3.6 s. Consequently, the number of payload bytes are typically limited, e.g. to 51 payload bytes for the SF modes 10, 11, and 12 in case of the European TTN parameter set. This corresponds to a maximum on-air time of approx. 3 s for SF12.

For mioty the curve shows a flat behavior for few input bytes. This is caused by the 24 core-bursts that have to be filled first. Eventually, extension bursts have to be added when the payload capacity of the core-bursts is exhausted. The overall maximum on-air time of mioty is approx. 4 s for on packet.

Figure 3 shows the spectral footprint of mioty and LoRa for different payload length $N_{Payload}$ and different LoRa modes. The y-axis is on logarithmic scale, as the differences between the different configurations are so huge. It becomes obvious that already the fastest LoRa mode SF7 has a spectral footprint that is 6.5 times bigger than the spectral footprint of mioty for a payload size of ten bytes. If we compare mioty and LoRa SF12 – which approximately have the same robustness – LoRa has an 150 times bigger footprint. This corresponds to a soccer field against the complete state of Monaco. We will come back to the spectral footprint when comparing the network capacity of both systems.

4 Laboratory Performance Measurements

We conducted extensive measurements to analyze the performance of real LoRa and mioty devices. A major challenge with both systems are their long on-air times. At the same time, a large number of packets is needed to obtain statistically meaningful results, leading to measurement durations of several hours to days for just a single parameter configuration. Since such durations are impractical with a manual setup, we developed an automated measurement system

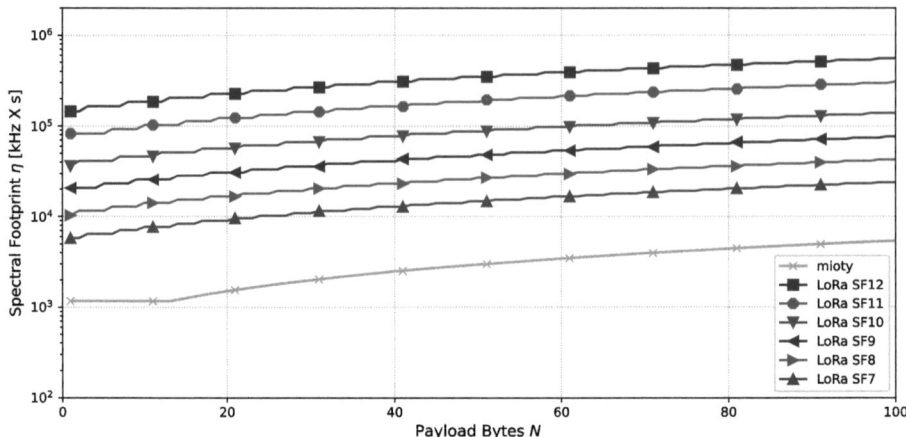

Fig. 3. Spectral footprint of mioty and different LoRa modes (European TTN parameter set) as a function of the number of payload bytes. Please note that the y-axis is scaled logarithmically.

capable of operating autonomously. Using this setup, the automated measurements took approximately three months in total.

The first subsection describes the hardware setup. Next, we present measurement results for sensitivity and capacity. Additional results related to mobility and robustness in license-exempt frequency bands under interference will be addressed in future publications.

4.1 Hardware Setup and Generation of the Test Signals

To enable automated measurements and to support scenarios involving many transmitters and parameter configurations, a Software-Defined Radio (SDR) setup was the only practical option for generating the required signals. Consequently, no real LoRa or mioty transmitters were used to produce the transmit data. We are fully aware that this may have an impact on the overall results, but there was no practical alternative—especially for the measurement of high traffic. Nevertheless, the results show good agreement with the values reported in the respective datasheets.

Figure 4 shows the hardware setup used for the laboratory measurements. A controlling PC generated the required data streams based on the configured transmission parameters. The SDR frontend, an Ettus USRP B210 [24], transmitted the data stream at a center frequency near 868 MHz—the typical LPWAN band in Europe. The B210 is equipped with a 12-bit digital-to-analog converter and supports a sampling rate of up to 61.44 Msps (mega-samples per second). To limit the received signal strength, multiple attenuators were used to reduce the signal power by 37 dB. The signal was then split into two coaxial lines using a power splitter, which introduced an additional attenuation of 3 dB. As a result,

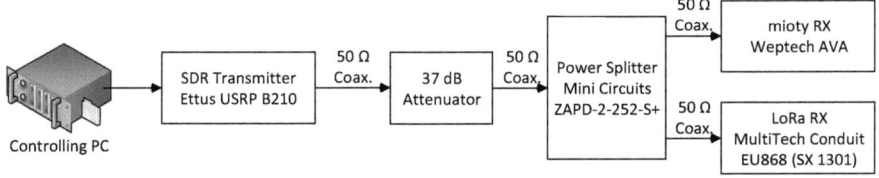

Fig. 4. Measurement setup used in the laboratory experiments. The mioty and LoRa receivers are both connected to the SDR transmitter. An attenuation of 40 dB was applied between the SDR front end and the receivers.

the total attenuation between the USRP B210 and the receivers was approximately 40 dB. This setup allowed both LoRa and mioty receivers to operate simultaneously. The mioty gateway used throughout the measurements was the Weptech AVA, running software version 4.9.2 [28]. It is based on a Raspberry Pi 4 combined with a transceiver module. This setup represents a relatively low-cost solution compared to other commercially available mioty base stations. The LoRa gateway used was the MultiTech Conduit EU868 [21], which is equipped with a Semtech SX1301 LoRa chip [3]. The gateway supports concurrent decoding of up to eight channels. Both gateways are available at a relatively low price, each well below 1000 €. The mioty and LoRa gateways were also connected to the controlling PC, which received the transmitted data along with additional metadata (e.g., signal-to-noise ratio). This allowed full automation of the measurement campaign, which ran continuously for approximately three months. The PC also handled automatic logging and analysis of the transmissions for subsequent evaluation.

The required SDR transmit signals were generated using an extended version of the open-source mioty project available on GitHub [26], which was enhanced to support LoRa signal generation. The implementation offers flexible transmission parameter configuration and includes channel models with integrated noise generation. All test data streams were generated at a sampling rate of approximately 3 Msps—significantly exceeding the bandwidths of mioty and LoRa waveforms. Due to the difficulty of precisely controlling reception levels in a software-based setup, additive white Gaussian noise (AWGN) was embedded directly in the generated signals to achieve the desired signal-to-noise ratio (SNR). As a result, the noise level was set significantly above the thermal noise floor. This approach simplifies system comparison by removing the impact of the frontend noise figure, allowing direct assessment of waveform performance. We consistently assume a noise level of -174 dBm/Hz, corresponding to the thermal noise floor and equivalent to a receiver with a 0 dB noise figure [23]. Given the 3 Msps sampling rate, AWGN occupies a 3 MHz bandwidth, which is much wider than the actual signal bandwidth and thus does not affect performance outcomes.

4.2 Sensitivity Measurements

We conducted measurements to evaluate the sensitivity of LoRa and mioty in an additive white Gaussian noise (AWGN) channel. This channel represents the simplest case, introducing only noise without any effects from multipath propagation or interference. Figure 5 shows the measurement results for both systems. Each configuration involves at least 1000 transmitted packets, allowing reliable estimation of packet error rates (PER) down to approximately 1% (i.e., 10^{-2}). Evaluating lower PERs would require a significantly larger number of transmissions, which is impractical, as the existing measurements already spanned several months.

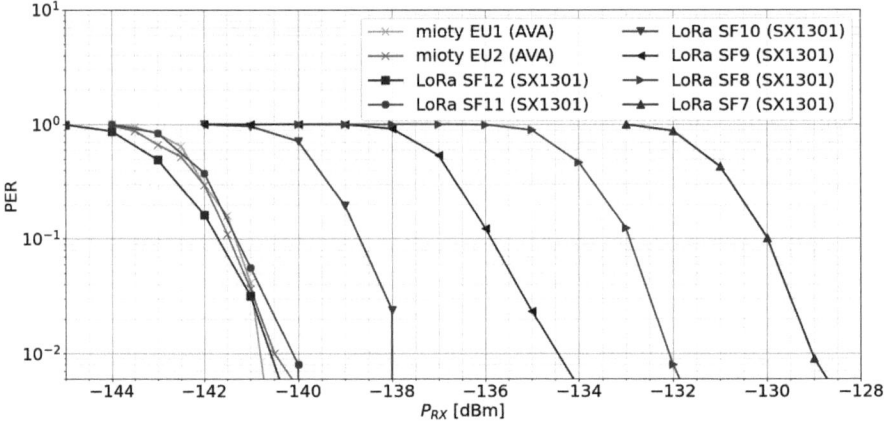

Fig. 5. Measurement results of mioty and LoRa in the Additive White Gaussian Noise (AWGN) channel. Increasing the SF by one results in a gain of approx. 3 dB. mioty has a comparable performance as LoRa with SF11 and SF12.

The long-range capability of LPWAN systems arises from their exceptionally high receiver sensitivities, typically around −142 dBm. Such sensitivity can only be achieved at very low payload bit rates. Consequently, the maximum theoretical bit rate at this sensitivity level is 2.2 kbps [25, Fig. 3], while in practice, due to various implementation losses, the achievable bit rate is considerably lower.

LoRa and mioty use different approaches to reach these low bit rates, and correspondingly, these low reception levels. LoRa employs spread-spectrum modulation, distributing bits over a wide bandwidth, whereas TS-UNB relies on ultra-narrowband frequency shift keying (FSK) combined with frequency hopping. LoRa's use of multiple spreading factors allows it to adapt sensitivity in steps of approximately 3 dB, where 3 dB corresponds to a doubling of power in linear terms. This tradeoff is reflected in both the transmit duration and the effective payload bit rate. In general, halving the bit rate yields a theoretical link budget improvement of 3 dB, a principle rooted in information theory and broadly applicable to communication systems in practice [22].

Table 1 summarizes the measurement results and compares them to both the data sheet sensitivities and the theoretical sensitivity bound. The latter represents the minimum required sensitivity of an ideal communication system to support the same payload bit rate, considering only thermal noise at $-174\,\text{dBm/Hz}$ [25]. This bound is based on the theoretical channel capacity, which is overly optimistic for the short packet lengths typical of LPWANs [14]. Nonetheless, it provides a useful reference for evaluating the performance gap of real systems. The achievable sensitivity refers to the lowest observed sensitivity at which successful decoding is possible using real hardware. The difference between this value and the theoretical bound represents the implementation loss. For mioty, this gap is approximately 6.5 dB, while for LoRa it ranges from 8 dB to 9 dB. Consequently, mioty slightly outperforms LoRa.

Table 1. Parameters for different system configurations assuming 10 bytes of application payload (all values are approximate values). The LoRa values are for a bandwidth of 125 kHz. The bound indicates the minimum required reception level in case of a non-existing perfect system. *) The datasheet values for LoRa are for a Packet Error Rate of 10%. All other values are for 1%.

Mode	Typical TX Duration	Theoretical Bound [25, (4)]	Datasheet Sensitivity*)	Measured Sensitivity
LoRa SF7	~62 ms	−138 dBm	−126.5 dBm [3, Table 12]	−129 dBm
LoRa SF8	~113 ms	−141 dBm	−129.0 dBm [3, Table 12]	−132 dBm
LoRa SF9	~206 ms	−143 dBm	−131.5 dBm [3, Table 12]	−134.5 dBm
LoRa SF10	~371 ms	−146 dBm	−134.0 dBm [3, Table 12]	−138 dBm
LoRa SF11	~823 ms	−148 dBm	−136.5 dBm [3, Table 12]	−140 dBm
LoRa SF12	~1483 ms	−151 dBm	−139.5 dBm [3, Table 12]	−140.5 dBm
mioty EU1	~360 ms (3.5 s)	−147 dBm	−138 dBm [9]	−140.5 dBm

4.3 Power Consumption

Energy measurements were conducted using the SX1276MB1MAS evaluation board[2], which is equipped with the SX1276 chip [7]. The transmit power was set to 13 dBm (20 mW). The measurement device was a Rohde & Schwarz HMC0815 power analyzer. Only the power consumption of the evaluation board – which did not include the controlling microprocessor – was measured. In order to increase the accuracy always ten packets were measured to calculate the energy per packet. The current during TX was approx. 38 mA, i.e., all modes (the LoRa and mioty) exactly had the very same current consumption.

Given the identical power consumption across modes, we can compute the energy per packet and, consequently, estimate battery lifetime. Table 2 presents theoretical values based on the SX1276 datasheet. It includes the maximum number of transmittable packets assuming a 1000 mAh battery at 3.3 V, as well

---

[2] https://www.semtech.com/products/wireless-rf/lora-connect/sx1276mb1mas, accessed: June 2025.

Table 2. Theoretical energy for transmitting one packet with 10 bytes payload with 13 dBm (20 mW) using the values given in the SX1276 datasheet [7] (electrical power consumption 95 mW at 3.3 V. All mioty modes have identical values.

	SF7	SF8	SF9	SF10	SF11	SF12	mioty
Theor. Energy/Pkt [mJ]	5.9	12.6	19.6	35	78	141	38.2
On-air Time [s]	0.062	0.133	0.206	0.371	0.823	1.483	0.403
Power Cons. during TX [mW]	95	95	95	95	95	95	95
Max. Packets 1000mAh Bat.	2 m	942 k	606 k	340 k	152 k	84 k	310 k
Lifetime (96 Pkts/Day) [years]	57	27	17	9.6	4.3	2.4	8.9

as the expected battery lifetime when sending one packet every 15 min (i.e., 96 packets per day). Note that these values do not account for power consumption from other system components, such as the microcontroller.

LoRa modes SF11 and SF12 offer robustness comparable to mioty in an AWGN channel. However, mioty achieves this robustness with significantly shorter on-air times. As a result, mioty provides much longer battery life under similar performance conditions. Lower spreading factors in LoRa consume less energy due to reduced transmission durations, but this comes at the cost of significantly lower robustness. This could be particularly beneficial in the case of ADR, which was not analyzed in this study. Overall, these results highlight mioty's closer alignment with the theoretical performance limits shown in Table 1.

4.4 Network Capacity Measurements

mioty and LoRa typically employ the ALOHA channel access scheme, originally developed to connect the Hawaiian islands in the 1970s [12]. In ALOHA, devices transmit data with minimal coordination, sending packets whenever they are ready. Notably, the widely used "Listen Before Talk" (LBT) mechanism is not applied. LBT requires a device to sense the channel and transmit only if it appears idle, which is unsuitable for LPWAN systems due to their extremely low signal levels. Since ALOHA offers no protection against simultaneous transmissions, the likelihood of packet collisions increases with the number of active devices and the duration of their transmissions. A comparison of the spectral footprint η of LoRa SF12 and mioty implies that mioty achieves a 150-fold higher capacity for a comparable system bandwidth. The detailed performance of both systems is analyzed in the following paragraphs.

Theoretical Analyses of LoRa: The capacity of LoRa networks and potential enhancements have been extensively studied in the literature (e.g., [13,15]). A key factor determining network capacity is the Spreading Factor (SF), which directly impacts the packet duration. Table 1 shows the transmit durations for different SFs, ranging from 7 to 12. In general, a higher SF improves robustness against noise but increases transmission time. For example, transmitting 10 bytes of payload with SF7 takes approximately 62 ms, whereas SF12 requires around 1483 ms. Due to its shorter transmission time and narrower spectral footprint, SF7 supports a significantly higher overall network capacity compared to SF12—but at the cost of requiring a substantially higher reception level.

Table 3. Theoretical capacity of a LoRa system for different channel access schemes in Pkts/min for one 125 kHz channel

	SF7	SF8	SF9	SF10	SF11	SF12
Fully Scheduled	968	531	291	162	73	40
Max. ALOHA	178	98	54	30	13.4	7.4
ALOHA 10% PER	51	28	15.3	8.5	3.8	2.1
ALOHA 1% PER	4.8	2.7	1.5	0.8	0.4	0.2

Table 3 lists the throughput of different parameter configurations and channel access schemes. The channel parameters are again the European TTN parameter set, assuming a channel bandwidth of 125 kHz [11]. A fully scheduled system would offer the maximum capacity. This means the channel is fully occupied, with no pauses between packets and no collisions. Generally, this would require a full coordination of the network, which is not feasible in LPWAN. Therefore, this is an upper bound on the achievable network capacity.

Practically more relevant is the ALOHA capacity [12]. It gives the maximum achievable throughput in case of fully random access and identical packet length. The maximum capacity is reached for a PER of 63%, which is most likely too high for most use-cases. More relevant is the payload throughput in case of lower PER. The resulting capacity is given by $C_{Max.PER} = -0.5 C_{FS} \ln(1-\text{PER})$ [12], where ln() stands for the natural logarithm.

The obtained capacity values may appear low, especially considering the large coverage area typical of LPWANs. However, they align with results reported in [17] and figures published by Semtech in their application notes. The SX1301 gateway chip can decode nine parallel LoRa channels [3], and "The Things Network" specifies eight uplink channels [11]. As a result, the capacity can be scaled by a factor of eight, assuming eight 125 kHz-wide channels are available, requiring a total bandwidth of 1 MHz.

Additionally, the SX1301 supports simultaneous reception of multiple spreading factors (SFs) on the same channel. This is enabled by the process gain [22] inherent to spread-spectrum modulation, such as that used by LoRa. The process gain significantly attenuates uncorrelated signals—such as those using different SFs—allowing concurrent reception. According to Semtech's documentation [10, Tab. 1], if signals using SF7 to SF12 are received with sufficiently distinct power levels, it is theoretically possible to decode all of them concurrently. In this case, the total per-channel capacity is the sum of the individual SF capacities, as listed in Table 3. However, in practice, this benefit is limited for weak sensor nodes operating with SF12, which contribute little to the overall capacity due to their low data rates and high on-air time.

Theoretical Analyses of Mioty: The splitting of jointly encoded data into multiple bursts, as used in mioty, precludes the use of the simple capacity estimation applied to LoRa. Instead, a more detailed analysis is required. An in-depth evaluation of packet splitting is provided in [18], where Fig. 10 illustrates the resulting packet error rate (PER) as a function of channel load.

mioty transmits each packet within at least 24 bursts. According to [18, Fig. 10], a PER of approximately 1% can be expected at a channel load of about 40%. It should be noted, however, that the following calculation is a rough approximation. In EU1 mode, two 60 kHz wide uplink frequency bands are used, resulting in a total bandwidth of $BW = 120$ kHz. Assuming a spectral load of 40% (corresponding to a 1% PER), the capacity for a single mioty gateway can be estimated as:

$$C_{\text{EU1, 1\%}} \approx \frac{BW}{\eta_{\text{mioty,real}}} \times \text{Load} = \frac{120\,\text{kHz}}{1.1687\,\text{kHz} \cdot \text{s}} \times 0.4 \approx 2578\,\text{Pkts/min} \quad (1)$$

A similar calculation can be performed for the wideband modes (EU2, US0), which use two 720 kHz wide bands. The resulting network capacity for a single gateway is approximately $C_{\text{EU2, 1\%}} \approx 30,936$ Pkts/min.

In low-latency modes, the effective on-air time remains unchanged. However, these modes use only a single time-frequency pattern, increasing the likelihood of full collisions between messages. In such cases, the assumptions made in [18] no longer hold, and a degradation in performance can be expected.

Laboratory Measurements: The primary challenge for the capacity characterization was the low capacity of LoRa, which led to extremely long measurement durations to achieve a relevant amount of received packets. Each individual measurement consists of a 30 s activation phase, a 60 s core phase, and a 30 s deactivation phase, totaling 120 s. Within this duration, data packets are randomly distributed according to a Poisson point process [27], which models uncoordinated transmissions—characteristic of ALOHA-based access schemes and typical for LPWAN networks. The core phase always spans exactly one minute and serves as the basis for analysis. In order to get statistically relevant measurement results, at least 20 core phases were measured for each parameter configuration, which already lasted in a measurement duration of 40 min.

The activation and deactivation phases were included to maintain a constant channel load during the core phase. Without them, the start and end of the measurement period would exhibit lower packet densities, potentially leading to overly optimistic results. Only packets that begin transmission within the core phase are counted, which is ensured by assigning different addresses to packets originating in the core versus the activation/deactivation phases. The number of packets transmitted in the core phase is fixed and corresponds exactly to the target number of packets, simplifying comparison across measurements. However, this adjustment breaks the strict Poisson property, introducing inaccuracies—particularly when very few packets per minute are transmitted. A correct modeling of the Poisson property would have further increased the measurement time. Figure 6 shows an example spectrogram of a mioty signal in wideband (EU2) mode with a packet rate of 2000 Pkts/min. The resulting PER for this density was practically zero.

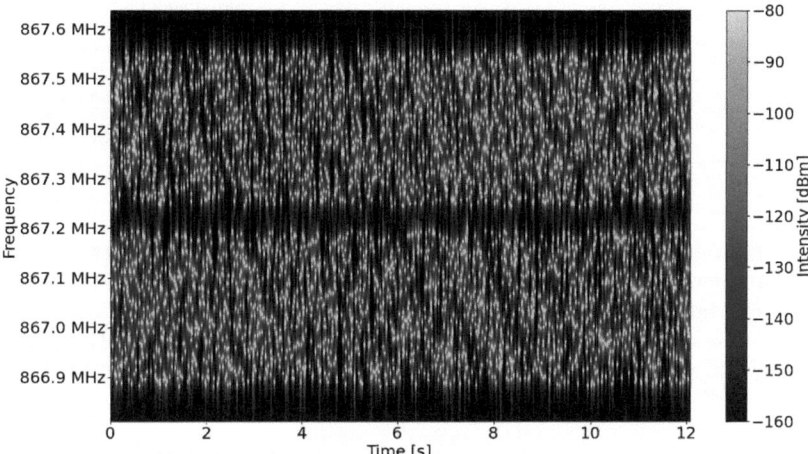

Fig. 6. Spectrogram of mioty with 2000 Pkts/min in wideband (EU2) mode. The so-called upper and lower bands are clearly visible.

The initial measurements assumed an identical reception level of -110 dBm for all packets, effectively neglecting the impact of thermal noise at the receiver. For LoRa, no frequency offset was considered, as it most likely has no relevant effect on the results. In contrast, for mioty, a frequency offset of up to 20 ppm was assumed—representing the typical tolerance of low-cost crystal oscillators. The offset was uniformly distributed within this range.

Figure 7 shows the measured packet error rate (PER) as a function of the number of packets per minute. The measured curves for both LoRa and mioty align well with the theoretical predictions. LoRa generally supports only a very limited number of packets per minute at an acceptable PER. As expected, reducing the SF by one approximately doubles the maximum packet rate for a given PER, consistent with the theoretical results in Table 3. In contrast, mioty's normal mode (EU1) can support approximately 6,000 Pkts/min at a PER of 1%, and over 15,000 Pkts/min at a PER of 10%. Interestingly, the performance gap between the normal (EU1) and wideband (EU2) modes is relatively small. Although the wideband mode uses twelve times more channel bandwidth, it does not exhibit a proportionally higher capacity. A possible explanation is that the AVA gateway software was not capable of handling such high traffic volumes. The low-latency modes exhibit a higher PER, likely due to the use of a single

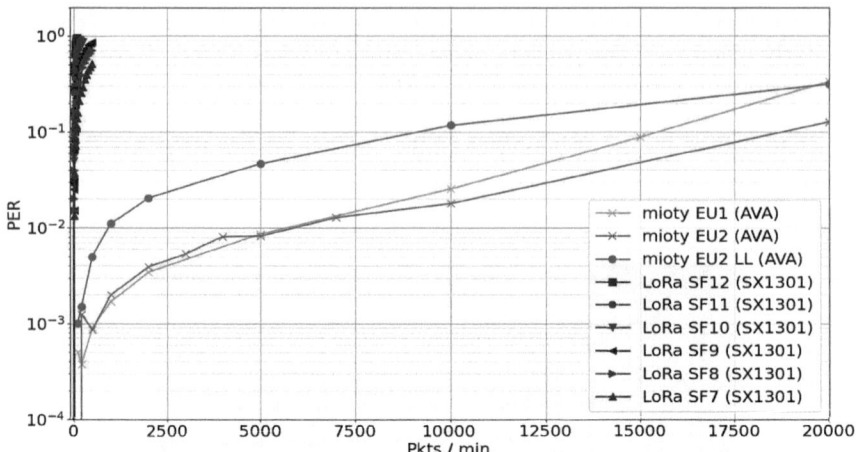

Fig. 7. Measurement results of mioty and LoRa as a function of packets per minute, assuming an identical reception level for all packets. At least 20 min of real-time data were used for the measurements. LoRa was measured using a single channel, resulting in a comparable total bandwidth of 125 kHz for both LoRa and mioty EU1. For typical LoRa gateways with eight channels, the resulting LoRa capacity has to be multiplied by eight.

frequency-hopping pattern, which increases the probability of collisions. Nevertheless, the measured results exceed the predictions made by the theoretical analysis.

Figure 8 provides a closer view of the LoRa modes using SF11 and SF12. The dashed lines indicate the theoretical PER based on ALOHA capacity calculations (cf. Table 3). At very low packet rates, the measurements exhibit significant fluctuations due to the small number of packets captured over just 20 core frames. At higher loads, however, the measured curves closely follow the theoretical ALOHA predictions [12]. Overall, these results highlight the low capacity of LoRa networks with high SF, where the obtained measurement results strictly follow the official Semtech application note AN1200.64 [8, Table 6]. The application note mentions a LoRa capacity of 5000 packets per day with SF12 and a 10% PER in a single 125 kHz channel, which is identical to 3.5 Pkts/min. This value matches closely with the measurement results depicted in Fig. 8 and the theoretical calculations in Table 3.

The previous measurements assumed a uniform distribution of reception levels. However, this is not realistic, as LPWAN devices are typically distributed over a wide geographic area, leading to varying distances from the gateway and, consequently, differing reception levels. For the following measurements, we adopt the node distribution model described in [27, Sec. IIIb]. The minimum and maximum distances between a device and the gateway are set to $r_{min} = 100$ m and $r_{max} = 5000$ m, respectively. A rural propagation environment is assumed, with the gateway antenna positioned at a height of 140 m. Under this model,

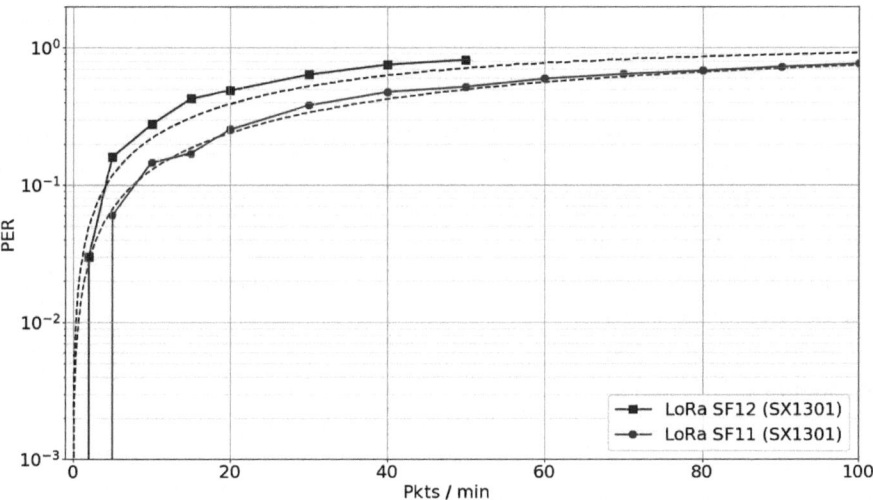

Fig. 8. Detailed analyses of LoRa for SF 11 and SF 12 as function of the packets per minute for a single channel. The dashed lines indicate the theoretical ALOHA PER. Accurate estimation of very low PER would necessitate ultra-long measurement durations.

deviations from the previous results are expected due to the so-called capture effect: when strong and weak signals arrive simultaneously, the gateway is able to decode the stronger one. However, the measurements have shown that there was only a small difference between both measurements.

Table 4 presents the measurement results for a packet error rate (PER) of 10%. The results are shown for a typical gateway configuration—eight channels for LoRa with the TTN parameters [11] using 125 kHz bandwidth—and are also normalized to a total bandwidth of 1 MHz. The comparison clearly demonstrates that mioty significantly outperforms LoRa. In particular, LoRa's capacity is highly constrained for higher SFs, even when using all eight channels, which already consume a substantial portion of the available spectrum. In networks operating near the maximum capacity, the addition of new LoRa networks—especially those outside the operator's control—can significantly increase the resulting packet error rate. While installing additional gateways may allow more devices to use lower SFs and thus increase capacity, this comes at additional cost and compromises the fundamental advantages of LPWAN deployments. This limitation also applies to interference from non-LPWAN devices, which are common in license-exempt frequency bands. Measurements—whose results will be detailed in a future publication—demonstrate LoRa's high sensitivity to such interference.

Table 4. Measured capacity with 10% packet error rate (PER) normalized to a bandwidth of 1 MHz, and for one typical gateway (8 channels LoRa, 1 channel mioty) assuming the realistic reception level distribution

Mode	Normalized to 1 MHz (Pkts/min)	Typical Gateway (Pkts/min)
mioty EU1 (AVA)	~44,000	~8,000
mioty EU2 (AVA)	~10,400	~15,000
mioty EU1 LL (AVA)	~15,000	~2,700
mioty EU2 LL (AVA)	~2,700	~4,000
LoRa SF7 (SX1301)	~600	~600
LoRa SF8 (SX1301)	~370	~370
LoRa SF9 (SX1301)	~190	~190
LoRa SF10 (SX1301)	~125	~125
LoRa SF11 (SX1301)	~80	~80
LoRa SF12 (SX1301)	~40	~40

5 Conclusions

This paper has provided a detailed comparison of LoRa and mioty, two prominent LPWAN technologies, based on theoretical analysis and extensive measurements. While LoRa is widely adopted and simple to deploy, our results reveal significant limitations in terms of capacity and scalability—especially when higher spreading factors are used. mioty, by contrast, demonstrates a much closer alignment with theoretical performance limits. Its telegram-splitting technique enables high resilience against collisions, allowing for considerably higher packet throughput even in dense network environments. mioty also offers notable advantages in energy efficiency and battery lifetime when operating at comparable levels of robustness. Although LoRa can achieve low energy consumption with lower spreading factors, this comes at the cost of reduced coverage and reliability.

Overall, mioty emerges as the more robust and scalable solution for modern IoT deployments, particularly in capacity-constrained environments. These findings provide valuable guidance for network operators and system designers in selecting the most suitable LPWAN technology for their application requirements.

Disclosure of Interests. The work behind this paper was financed by the mioty Alliance. However, this had no impact on the results of this study.

Appendix – Calculation of the On-Air-Times

mioty

The physical layer is able to transmit up to 186 bits in the so-called core frame. This corresponds to 13 payload bytes for a typical configuration (Fixed MAC, short address mode). If less payload data are transmitted, the data is stuffed to fill at least one core frame, which covers 24 radio bursts. Each radio burst further consists of 24 forward error-corrected payload bits and a 12 bit mid-preamble for synchronization. This results in a radio burst time of 15.1 ms for the given symbol rate of 2380.371 Sym/s. If the data length exceeds the 186 bits (or 13 payload bytes), so-called extension frames – i.e. additional bursts – are added, increasing the overall on-air-time.

Each extension burst carries 8 payload bits. After convolutional encoding, this corresponds to 24 MSK symbols. In addition, a 12-bit midamble is inserted. Consequently, each extension hop increases the packet duration by 15.1 ms. The total on-air time of a mioty transmission for the aforementioned parameter configuration is hence given by:

$$T_{\text{mioty,ideal}} = 363\,\text{ms} + \max(0, N_{Payload} - 13) \times 15.1\,\text{ms} \qquad (2)$$

where $N_{Payload}$ is the number of payload bytes. The overall on-air time is independent of the actual mode, as the mode only adapts the frequency hopping pattern and the pause between the individual bursts.

However, in a practical system the time given in Eq. (2) is typically too optimistic. The MSK modulation requires predecessor symbols before and successor symbols after the actual burst. In addition to the ramp-up and ramp-down time this results in approx. 4 additional symbols that are typically added. Consequently, each radio burst has a realistic duration of 16.8 ms. Thus, the actual on-air time of a mioty transmission is approximately:

$$T_{\text{mioty,real}} = 403\,\text{ms} + \max(0, N_{Payload} - 13) \times 16.8\,\text{ms} \qquad (3)$$

The on-air time is only one parameter. For a complete network the spectral footprint is more relevant. The bandwidth of mioty can be approximated by 2.9 kHz [22]. Hence, we can approximate the spectral footprint of mioty by:

$$\eta_{\text{mioty,real}} = 2.9\,\text{kHz} \times T_{\text{mioty,real}}. \qquad (4)$$

As the different modes only affect the pauses and the hopping pattern, this spectral footprint is identical for all modes.

LoRa

The main parameters that affect the on-air time are the bandwidth BW, the spreading factor SF, and the data length in bytes N. The overall on-air time is given by[3]:

$$T_{LoRa} = T_{preamble} + T_{payload} \qquad (5)$$

with the symbol rate

$$T_{sym} = \frac{2^{SF}}{BW}$$

---
[3] https://www.rfwireless-world.com/calculators/LoRaWAN-Airtime-calculator.html, accessed: June 2025.

the preamble duration

$$T_{preamble} = (N_{preamble} + 4.25) \times T_{sym}$$

and the duration of the payload part

$$T_{payload} = \left(8 + \max\left(\left\lceil \frac{8N - 4SF + 28 + 16 - 20H}{4(SF - 2DE)} \right\rceil \times (CR + 4), 0\right)\right) \times T_{sym}$$

where $\lceil \; \rceil$ is the ceil operator.

H indicates the presence of a header (present if $H = 0$, not present if $H = 1$), $N_{preamble}$ is the preamble length in chirps, and CR defines the code-rate of the forward error-correction (code-rate 1 (i.e. no FEC) for $CR = 0$, code-rate 4/5 for $CR = 1$, code-rate 2/3 for $CR = 2$, code-rate 4/7 for $CR = 3$, code-rate 1/2 for $CR = 4$). Finally, DE is the low data-rate optimization mode (disabled if $DE = 0$, enabled if $DE = 1$). The ceil operator results in a non-linear behavior between the data length N and the on-air time. This is caused by the internal interleaving working on fixed length data blocks.

Typical parameters for Europe are provided, for example, by the frequency plan of The Things Network (TTN) [11], a community-driven LoRaWAN network: Bandwidth $BW = 125$ kHz; preamble length $N_{preamble} = 8$; coderate of the forward error correction of 4/5, i.e. $CR = 1$; header in explicit mode, i.e. $H = 0$; low data-rate optimization mode enable for SF11 and SF12 for 125 kHz (i.e. $DE = 1$), and disabled for the other SF (i.e. $DE = 0$). Using these parameters results in the following on-air-time:

$$T_{LoRa,TTN} = \left(20.25 + 5\left\lceil \frac{2N - SF + 11}{(SF - 2DE)} \right\rceil\right) \times T_{sym} \quad (6)$$

However, this equation does not consider the overhead due to the LoRaWAN protocol, which is 13 bytes for a normal payload packet [4]. Consequently, we get $N = 13 + N_{Payload}$, where $N_{Payload}$ is the actual payload length in bytes.

We can again calculate the spectral footprint. For LoRa using the TTN parameters it is given by:

$$\eta_{\text{LoRa,TTN}} = BW \times T_{LoRa,TTN} = \left(20.25 + 5\left\lceil \frac{2N - SF + 11}{(SF - 2DE)} \right\rceil\right) \times 2^{SF} \quad (7)$$

Interestingly, the spectral footprint of LoRa does not depend on the bandwidth of the signal, but is mainly affected by the spreading factor.

References

1. Advanced encryption standard (AES) (2001). https://nvlpubs.nist.gov/nistpubs/FIPS/NIST.FIPS.197.pdf
2. IEEE standard for low-rate wireless networks. IEEE Std 802.15.4-2015 (2015). https://standards.ieee.org/standard/802_15_4-2015.html

3. SX1301 datasheet. Revision 2.0 (2017). https://www.semtech.com/products/wireless-rf/lora-gateways/sx1301. Accessed June 2025
4. LoRaWAN specification. LoRa alliance specification 1.0.3 (2018). https://lora-alliance.org/sites/default/files/2018-07/lorawan1.0.3.pdf. Accessed June 2025
5. Short range devices; low throughput networks (LTN); protocols for radio interface A. ETSI TS 103 357 V1.1.1 (2018). https://www.etsi.org/deliver/etsi_ts/103300_103399/103357/01.01.01_60/ts_103357v010101p.pdf. Accessed June 2025
6. Short range devices (SRD) operating in the frequency range 25 MHz to 1 000 MHz; part 2: Harmonised standard for access to radio spectrum for non-specific radio equipment. ETSI EN 300 220-2 V3.2.1 (2018). https://www.etsi.org/deliver/etsi_en/300200_300299/30022002/03.02.01_60/en_30022002v030201p.pdf. Accessed June 2025
7. SX1276/77/78/79 datasheet. Revision 6 (2019). https://www.semtech.com/products/wireless-rf/lora-transceivers/sx1276. Accessed June 2025
8. Application note AN1200.64: LR-FHSS system performance (2022). https://www.semtech.com/products/wireless-rf/lora-connect/sx1261. Accessed June 2025
9. IoT für kritische Infrastrukturen (2023). https://www.swissphone.com/de-at/aktuelles/news/iot-fur-kritische-infrastrukturen/. Accessed June 2025
10. Predicting LoRaWAN capacity (2023). https://lora-developers.semtech.com/documentation/tech-papers-and-guides/predicting-lorawan-capacity/. Accessed June 2025
11. The things network frequency plans (2023). https://www.thethingsnetwork.org/docs/lorawan/frequency-plans/
12. Abramson, N.: The ALOHA system: another alternative for computer communications. In: Proceedings of the Fall Joint Computer Conference. 17-19 November 1970, pp. 281–285 (1970)
13. Bankov, D., Khorov, E., Lyakhov, A.: Mathematical model of Lorawan channel access with capture effect. In: 2017 IEEE 28th Annual International Symposium on Personal, Indoor, and Mobile Radio Communications (PIMRC). IEEE (2017)
14. Dolinar, S., Divsalar, D., Pollara, F.: Code performance as a function of block size. TMO Progress Report **42**(133) (1998)
15. Elshabrawy, T., Robert, J.: Capacity planning of LoRa networks with joint noise-limited and interference-limited coverage considerations. IEEE Sens. J. **19**(11), 4340–4348 (2019)
16. Ferré, G., Giremus, A.: LoRa physical layer principle and performance analysis. In: 2018 25th IEEE International Conference on Electronics, Circuits and Systems (ICECS), pp. 65–68 (2018)
17. Furtado, A., Pacheco, J., Oliveira, R.: PHY/MAC uplink performance of LoRa class a networks. IEEE Internet Things J. **7**(7), 6528–6538 (2020)
18. Kilian, G., et al.: Increasing transmission reliability for telemetry systems using telegram splitting. IEEE Trans. Commun. **63**(3), 949–961 (2015)
19. Kilian, G., et al.: Improved coverage for low-power telemetry systems using telegram splitting. In: Smart SysTech 2013; European Conference on Smart Objects, Systems and Technologies (2013)
20. Molisch, A.F.: Wireless Communications. Wiley (2012)
21. MultiTech: MultiTech conduit EU868 datasheet (2023). https://www.multitech.com/documents/publications/data-sheets/86002217.pdf. Accessed June 2025
22. Proakis, J.G., Salehi, M.: Digital Communications, vol. 4. McGraw-hill New York (2001)
23. Rappaport, T.S., et al.: Wireless Communications: Principles and Practice, vol. 2. Prentice hall PTR New Jersey (1996)

24. Research, E.: Ettus B210 datasheet (2023). https://www.ettus.com/all-products/ub210-kit/. Accessed June 2025
25. Robert, J., Heuberger, A.: LPWAN downlink using broadcast transmitters. In: 2017 IEEE International Symposium on Broadband Multimedia Systems and Broadcasting (BMSB) (2017)
26. Robert, J., Neumueller, C.: TS-UNB-Lib (2023). https://github.com/mioty-iot/. Accessed June 2025
27. Robert, J., Rauh, S., Lieske, H., Heuberger, A.: IEEE 802.15 low power wide area network (LPWAN) PHY interference model. In: 2018 IEEE International Conference on Communications (ICC). IEEE (2018)
28. Weptech: Weptech AVA gateway datasheet (2023). https://www.weptech.de/de/produkte-funkloesungen/mioty-gateway-ava.html?file=files/site/pdf/datenblaetter-voegel/datasheet_ava_mioty-gateway_v1-0_samplekit.pdf. Accessed June 2025
29. Wuerll, R., Robert, J., Kilian, G., Heuberger, G.: Optimal one-shot detection of preambles with frequency offset. In: 2018 IEEE International Conference on Advanced Networks and Telecommunications Systems (ANTS) (2018)

Open Access This chapter is licensed under the terms of the Creative Commons Attribution 4.0 International License (http://creativecommons.org/licenses/by/4.0/), which permits use, sharing, adaptation, distribution and reproduction in any medium or format, as long as you give appropriate credit to the original author(s) and the source, provide a link to the Creative Commons license and indicate if changes were made.

The images or other third party material in this chapter are included in the chapter's Creative Commons license, unless indicated otherwise in a credit line to the material. If material is not included in the chapter's Creative Commons license and your intended use is not permitted by statutory regulation or exceeds the permitted use, you will need to obtain permission directly from the copyright holder.

NeuroKey: A Lightweight AI Tool Using Passive Keystroke Dynamics for Parkinson's Disease Detection and Monitoring

Ritu Chauhan[1], Mehak Jena[1], and Dhananjay Singh[2(✉)]

[1] Artificial Intelligence and IoT Lab, Centre for Computational Bio. and Bioinformatics, Amity University, Noida, India
[2] The Pennsylvania State University, University Park, USA
dsingh@psu.edu

Abstract. Parkinson's Disease (PD), a progressive neurodegenerative condition, sometimes eludes early diagnosis owing to its subtle and diverse motor symptoms. Conventional diagnostic methods are expensive, intrusive, and often unattainable. This paper presents the NeuroKey Prediction Tool, a lightweight, browser-based AI system that utilises passive keystroke dynamics gathered during regular typing to detect and track Parkinson's disease. Utilising a modular Streamlit framework, NeuroKey extracts statistical and temporal biomarkers, including key hold time and inter-key intervals, employing machine learning and deep learning models for classification and regression applications. The Random Forest classifier attained an accuracy of 88% in Parkinson's Disease detection, surpassing XGBoost (65%) and Logistic Regression (47%). In terms of predictive performance for motor severity estimation (UPDRS/nQi scores), Ridge Regression ($R^2 = 0.75$), LSTM (0.74), and Random Forest Regressor (0.69) exhibited robust results. These findings validate the feasibility of digital phenotyping using typing behavior as an effective, non-invasive biomarker for neuromotor evaluation. NeuroKey's lightweight, scalable, and privacy-conscious architecture facilitates home-based ambient assisted living and telemedicine processes, thereby improving proactive, patient-centered treatment. Future endeavors will enhance functionality by utilizing varied, real-world typing data, incorporating personalized baselines, and investigating federated learning to protect privacy while augmenting performance. NeuroKey showcases the integration of AI and IoT for accessible, continuous, and precise neurological healthcare.

Keywords: e-Health Monitoring · Parkinson's Disease · Keystroke Dynamics · Machine Learning · Deep learning · UPDRS-III · IoT Applications · Predictive Modelling

1 Introduction

The intersection of the healthcare industry with the fast-developing digital technologies has generated a paradigm shift into the proactive, patient-centered, and continuous healthcare models [1]. These digital-based healthcare innovations are collectively recognized as e-Health, and they cover everything in between telemedicine and wearable monitoring to AI-based diagnostics and personal treatment systems. One of the crucial subsets of e-Health is assisted living that connects with the increasing demand for autonomy, safety, and tele-care among elderly people, and particularly disabled and chronically ill individuals (Parkinson, etc.) [2]. The ubiquitous computing, the Internet of Things (IoT) sensor, and smart diagnostic tools trend in the contemporary world is helping healthcare systems to evolve beyond the constraints of hospital-based care and taking the diagnostic implications to the home and community setting. The Internet of Things (IoT) is driving this transformation forward as a distributed system of devices able to gather, process, and share health data in real time [3]. The use of IoT in devices such as smartwatches, biosignals, ambient sensors, and connected keyboards can become passive or active channels of transmission of vital health metrics [4]. There exists a capacity constraint known as translating that kind of low-level sensor information into useful, understandable insight, requiring clinical intervention every step of the way [5].

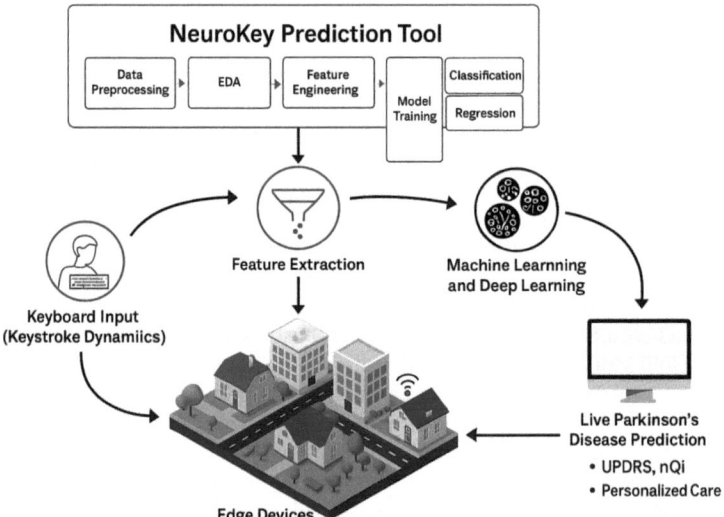

Fig. 1. An overview of NeuroKey system using keystroke dynamics and edge-enabled IoT infrastructure.

In the Fig. 1 depicts NeuroKey, an IoT-based monitoring system for Parkinson's disease. Keyboard inputs record keystroke dynamics, which are subjected to data preprocessing, exploratory analysis, and feature engineering. Machine and deep learning algorithms subsequently evaluate these features to forecast the severity of Parkinson's in real time on edge devices, facilitating personalized, non-invasive, ongoing therapy. Therefore the synthesis of an interface named NeuroKey Prediction Tool support minimal knowledge, is user-friendly, and easy-to-interpret Parkinson disease diagnosis (PD) prediction framework during its initial stages, which is a degenerative nerve disorder that affects motor ability and fine movements. Our tool can document minor motor symptoms long before they materialize into clinical conditions by examining time characteristics of typing (e.g. key hold time, inter-key timings). This renders the tool appropriate to be used in cases of home-based screening, tele-neurology support, and ambient assisted living in which conventional clinical tools may not be present. With the prevailing global trend of increased neurodegenerative disease onset, the need to deliver scalable systems which facilitate long-lasting, non-invasive patient monitoring procedures is at an all-time high [6]. Many studies indicate that the trend of neurological disorders, such as PD, is becoming extremely high with the aging societies [7]. Our research as NeuroKey Prediction Tool do not only offer early diagnostics but may also be used to monitor the disease progression via longitudinal typing patterns.

The core of many applications such as NeuroKey Prediction Tool involves Artificial Intelligence (AI) namely Machine Learning (ML) and Deep Learning (DL) and the integration of such models with medical applications are on the rise in automating the classification, prediction, and decision-making practices [8]. In this research, the Logistic Regression, Random Forest and XGBoost ML models have been applied in the classification of the extracted keystroke features in order to determine whether a person is a PD-positive or a healthy control. To provide more detailed evaluations, regression models provide estimates of clinical scale, such as UPDRS-III (Unified Parkinson's Disease Rating Scale) and nQi (neuroQWERTY index) that can be used to support binary screening as well as ongoing measure of severity [9]. The implementation of Deep Learning, namely LSTM networks, enables to model the time dependencies in typing sequence, thus, reflecting more complex neuromotor variability [10]. The combination of the ML and DL also enhances the versatility and flexibility of the tool to be used in different applications [11].

Our study leverages the forecasting of the Unified Parkinson's Disease Rating Scale (UPDRS-III) motor score, a widely-used measure in evaluation of the severity of motor symptoms of Parkinsonian patients by neurologists. Estimating UPDRS remotely within the processes of e-Health and assisted living is a form of scale capability to the in-clinic standard diagnosis that is resource-demanding and time-consuming, subjective, and typically inaccessible to the patient in the rural area or in the resource-constrained environment.

Our study proposes the NeuroKey Prediction Tool, which can be used to predict Parkinson's Disease occurrence and symptom severity through the use of keystroke dynamics. The tool has four tabs that are interactive to load the data of participants, perform analysis at cohort level, train machine and deep learning models, and provide prediction on real time input of typing. It can both classify the status of PD and regress the values of UPDRS-III and nQi scores using features based on key holding durations and time between the keys. Due to its use of typical keyboard input and interpretable AI models, the tool shows how non-invasive digital biomarkers have the potential to be user-friendly, accessible, and help screen people at an early age and across a great distance.

2 Literature Review

Parkinson's Disease (PD) is a degenerative neurodegenerative disease, whose main hindrance is the motor-related functions that occur after the destruction of dopamine neurons in the area of substantia nigra in the brain [12]. The Unified Parkinson s Disease Rating Scale (UPDRS) especially Part III which assesses the motor aspects is commonly used in the clinical setup to measure disease severity. It is a 0–108 scoring system (0 being no motor symptoms and 108 being severe motor loss) that has a unified value of measurement in terms of diagnosis and disease progression [13].

E-health systems are an evolving concept with a broad meaning such as digital apps and services aimed to optimise healthcare delivery, patient monitoring, and clinical decision support [16]. Assisted living systems are beginning to play more and more roles in decentralized at-home patient monitoring of patients with neurological, cardiovascular, or musculoskeletal pathologies. These systems also minimize the number of hospital visits, they minimize burden on care-givers, and also empower patients to participate in self-management [14]. Parkinson's Disease because of its chronic condition and course of its progression offers a perfect fit when it comes to being embedded into supportive assisted living environments by means of e-health health platforms [15].

Internet of Things (IoT) has been revolutionary towards providing real-time, ubiquitous sensing, communication and processing of the information relative to healthcare industry [16]. Health monitoring systems that rely on IoT are interconnected into a set of devices equipped with smartwatches, motion sensors, wearables, and ambient sensors which collect physiological, and behavioral data [17]. Then these data flows are either performed locally (at the edge) or delivered to the cloud platforms where they can be processed centrally. In PD and other motor disorders the IoT system has been applied in monitoring tremors, gait, balance, and more recently, typing dynamics [18]. The use of IoT in assisted living also contributes to the paradigm of ambient intelligence where the environment reacts to the health condition of a person without requiring them to interrupt their daily processes and activities in any way [19]. The trends are in line with the innovation of lightweight non-invasive platforms that use daily tools (keyboards or mobile phones) to offer a medical-level monitoring service.

Artificial intelligence (AI) as well as its sub categories of Machine Learning (ML) and Deep Learning (DL) have become pivotal in the areas of medical diagnostics, health informatics and personalized medicine in recent years [20]. The purpose of the ML algorithms is to seek regularities in large data sets, label sets clinical conditions, and forecast patient outcomes in a somewhat automated process and at a higher accuracy level than the use of conventional statistics [21]. ML models have been used in Parkinson studies on wearable sensor data, voice data, gait, and more recently keystroke dynamics.

Keystroke dynamics were further explored as non-invasive digital biomarkers for neurodegenerative and neuropsychiatric diseases. The previous studies showed the potential for identifying early Parkinson's disease (PD) motor symptoms using a single keystroke feature hold time while typing naturally on a computer with an AUC of 0.81, outperforming typical motor tests like alternating finger tapping (AUC = 0.75) and single key tapping (AUC = 0.61) [9]. Although the study was to be commended for passive, hardware-independent data gathering, dependence on a single biomarker restricted the characterization of motor function richness. Similarly, other studies, meta-analyzed 41 studies (3,791 PD, 254 MCI, 374 psychiatric patients), which gave pooled PD sensitivity and specificity of 0.86 and 0.83, respectively, and emphasized the potential of naturalistic data and multimodal deep learning techniques [23]. But the work provided meta-level evidence without executing or comparing particular algorithms with different conditions in real-world scenarios. Our research takes both bases forward by progressing from single-feature, single-model to a standardized multi-feature framework with inter-key intervals, flight durations, typing rate, and variability measures, tested on two heterogeneous datasets (Tappy and neuroQWERTY) in both binary classification and regression contexts (UPDRS-III, nQi score prediction). Our work utilize a hybrid modeling approach that combines classical machine learning (Random Forest, XGBoost) and deep learning (MLP, LSTM, Transformer) and can be embedded within a real-time, interactive Streamlit environment that supports retrospective and live PD evaluation. The multi-feature, multi-dataset, and multi-model pipeline not only enhances accuracy and robustness in diverse environments but also provides direct clinical score prediction and a deployable diagnostic tool.

3 Methodology

3.1 Principle of NeuroKey Prediction Tool

NeuroKey Prediction Tool is a proposed light weight diagnosis assistant which translates typing raw data into clinically meaningful interpretations. The essence of the idea is based on the fact that motor symptoms of Parkinson's Disease are known to appear in fine motor control i.e. mainly finger dexterity and timing regularity. A seemingly everyday task such as typing, the use of which by most people on digital devices may supply passive but dense stream of motor data without the requirement of external sensors. NeuroKey Prediction Tool is guided by the assumption that alteration of typing dynamics (hold time, inter-key intervals and typing variability) can be utilized not only to differentiate between Parkinson patients and healthy subjects, but also to quantify the level of symptoms in patients through regression models. This idea coincides with the new and promising sphere of digital biomarkers. In contrast to conventional clinic-based assessments in which a visit to a hospital and scoring on a neurological scale would be needed, NeuroKey Prediction Tool makes it more available, as only an internet browser and a keyboard are needed. It has the principle to provide non-invasive, privacy preserving and real-time e-health solution, which particularly would be useful in remote screening, telemedicine and longitudinal follow up.

3.2 Dataset Description

The work employs the neuroQWERTY project MIT-CS1PD and MIT-CS2PD data sets; which is a validated clinical benchmark at examining typing conduct in Parkinsonian Illness, or PD, experiments sourced from Physionet [22]. The data were gathered during the standardized conditions of experiments, during which a textual type-in was used with the respondents (both PD-diagnosed and healthy controls) typing in predetermined test-strings. Every typing spur has been recorded with a very high temporal resolution providing the exact timing of keystrokes given in nanoseconds, and this quality of data is especially advantageous toward locating subtle motor deficits of PD. The accompanying metadata of each subject comprises a unique participant ID (pID), the ground-truth PD diagnosis (gt_pd), the scale of motor symptoms according to the Unified Parkinson dose

Rating Scale - III (UPDRS-III), the proprietary score of the motor symptom free of Parkinson dose (nqScore), the score based on keystroke variations, and typingSpeed (words per minute) and a tapTest result (a finger tapping test). The typing sessions are recorded in different CSV files and contain the following keystroke properties: key identity, hold_time, press_time, release_time, which are all calculated as the number of seconds elapsed since the start of the typing activity. A total of 85 participants were utilized in the data set maintaining clinical variability and a balanced distribution in genders, age, and diagnosis categories.

3.3 Model Architecture

NeuroKey Prediction Tool adheres to an architecture pattern of horizontally scaled and modular design since it reflects the requirements of IoT-grounded health monitoring devices, which profile on scalability, human-oriented user interface, and real-time analysis. Our study implements the tool in an interactive, 4-tab Streamlit dashboard helping users navigate through a well-established diagnostic pipeline starting at the data exploration of participants to the training of the model that predicts Parkinson's Disease (PD) in real-time. The tool receives alpha-numerics key stroke data through a text box input and uses the typing data to facilitate feature extraction. The feature extraction gives attributes like hold and inter-key periods. After selecting the input, the machine will gather features. These features are pre-normalized expanded with a polynomial, and filtered for outliers, after that they are presented to a model of binary classification which can be built based on algorithms to predict the presence of PD. When PD is identified, regression models are activated, which estimates a standard clinical variable of the severity of the disease, the UPDRS-III motor score and to predict the neuroQWERTY index (nQi), a typing-based index of motor disability. Both features are regressors trained with power-transformed target and polynomial expansion approach to make the models accurate and post-process using the clinical value so as to get interpretable clinical values. Figure 2 shows NeuroKey's IoT platform captures keyboard typing patterns, processes and extracts information, and employs machine learning models to forecast the severity of Parkinson's disease in real time on edge devices. All the regular typing processes on keyboards of homes, offices, or intelligent public areas are employed as silent but strong digital biomarkers. Any device that is connected be it a PC, intelligent keyboard, or mobile typing interface recovers keystroke dynamics and sends these lightweight data streams to a local or edge processing system. After analysis, the outputs are represented in real-time dashboards, presenting live health evaluation that may be accessed remotely by clinicians or incorporated into tele-health platforms.

NeuroKey Prediction Tool can facilitate an end-to-end digital screening process due to its modular design, suitable to input, feature extraction, classification and regression tasks, creating a seamless e-Health application.

3.4 Model Training Pipeline

The NeuroKey Prediction Tool offers a lightweight, modular diagnostic pipeline for Parkinson's Disease based on keystroke dynamics. Its dashboard coordinates four primary functions: access to metadata, cohort-level EDA, model training, and real-time

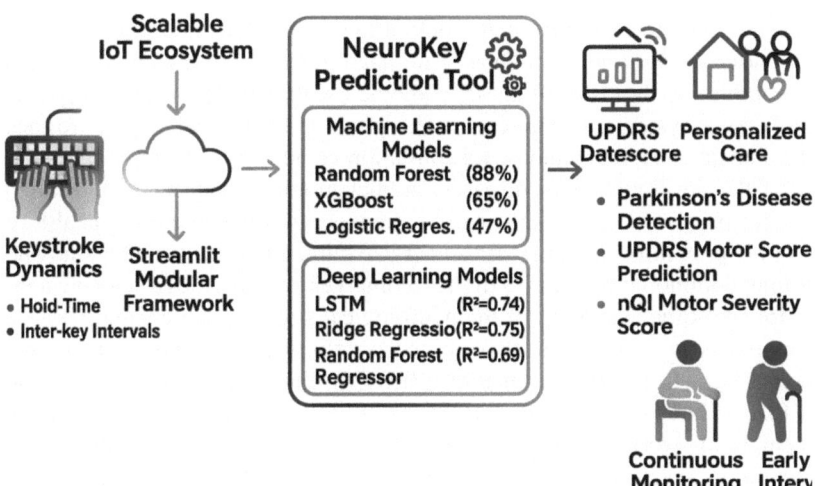

Fig. 2. Neuro Key Prediction Tool for Parkinson's Disease using IoT devices.

prediction. EDA shows that PD subjects type more slowly and erratically, with attributes such as hold time and inter-key interval having significant correlations with UPDRS and nQi scores. The pipeline of feature extraction calculates statistical features, enriches them with polynomial features, scales, performs power transforms for asymmetric targets, and detects outliers with Isolation Forests. These enriched, device-independent features enable classification (Logistic Regression, Random Forest, XGBoost) and regression (Random Forest Regressor, RidgeCV, ElasticNetCV, optional LSTM) models for reliable inference of PD status and motor severity. On the live prediction tab, on-the-fly feature extraction through typing is performed, and pre-trained models make instant predictions of PD presence, UPDRS-III, and nQi scores to provide real-time clinical insight with interpretability.

3.5 Tool Design and IoT e-Health Monitoring Integration

The NeuroKey Prediction Tool is a web-based, interactive dashboard constructed with Streamlit intended for Parkinson's Disease (PD) early detection and monitoring via keystroke dynamics. The tool has the minimum requirement of Python ≥ 3.8 and popular ML libraries (scikit-learn, XGBoost, TensorFlow, etc.) and can run on standard desktops and browsers with structured keystroke logs data. It extracts statistical features such as hold time and inter-key interval and uses trained machine learning and deep learning models to classify PD status and predict UPDRS-III and nQi scores. Subserving the objectives of e-Health and IoT-based health monitoring, NeuroKey facilitates continuous, non-invasive motor assessment with only a keyboard—obviating the requirement of specialized hardware. Its modular design facilitates integration with wearables, cloud-based EHR systems, and federated learning, and thus, it fits perfectly for usage in ambient assisted living (AAL), tele-neurology, and tailored remote care. The system could be placed right into the equipment of the households: ordinary keyboards, smart

TVs, or even voice-to-text apps, and would allow continuous neuromotor monitoring in quiet conditions. Follow-ups might be initiated due to alarming about subtle deteriorations stimulating the work of alerts, combined with home hubs or family caregiver dashboards. Such systems are applicable in hospitals and clinics when they are checking-in kiosks, bedside terminals and or rehabilitation centers, and they are able to help in monitoring the development of diseases, predict motor impairments, and individualize treatment programs. The results can be mapped with the electronic health records (EHRs) or hospital information systems to create clinician oversight. Our study helps in transforming common typing into a digital biomarker and provides a scalable and green solution for real-time screening of the neurological status in the contemporary digital health environment. At the present stage, the system was applied to publicly available Parkinson's datasets (Tappy and neuroQWERTY), including adult subjects with and without PD. Although the current study aims at verifying the tool within a controlled experiment, future work will involve integrating the IoT on a mass scale into the everyday home and clinical platforms in order to carry out neuromotor monitoring continuously and in real-time.

4 Results

4.1 Dataset Preprocessing and EDA

Their controlled clinical setting-derived dataset was parsed and used to retrieve high resolution keystroke sessions and aligned metadata. All typing sessions had millisecond-resolution dwell timestamps on key presses and releases enabling precise calculations of hold times and the inter key timings (IKIs). So during the preprocessing those sessions containing less than 10 valid keystrokes were dropped. Speed of typing, uniqueness of key usages and variability in timing were calculated with respect to corresponding sessions. Isolation Forests were used to identify and filter outliers to make sure that model training was not affected by noise. The continuous variables including the scores of UPDRS-III and nQi values were power-transformed to narrow down the skewness and the feature vectors were scaled by means of standard scaling. The input space was also enhanced by polynomial feature expansion as regression tasks.

UPDRS-III Score Distribution illustrated in Fig. 3 shows how often scores in Unified Parkinson's Disease Rating Scale Part III (UPDRS-III) are observed among the 85 subjects in the neuroQWERTY MIT-CSXPD dataset. The scores were ranged between 0 and 83 with the highest frequency of about 40 subjects at the lowest score (healthy controls), and another high density around the 29th score, where there also is a frequency of about 20 subjects.

A feature importance plot produced by the Random Forest Classifier as shown in Fig. 4, emphasizes the importance of chosen features (ht_std, ht_mean, Iki_mean, unique_keys, iki_std) to the prediction of Parkinson's Disease (PD) status. The plot reveals that the highest importance rating of ht_std (standard deviation of hold times) is nearly 0.30 and this indicates its high significance with the variability in typing behavior among PD subjects.

The heatmap indicates a positive relationship with a high value of correlation coefficient in ht_mean and ht_std with 0.605847, which suggests that the higher the average

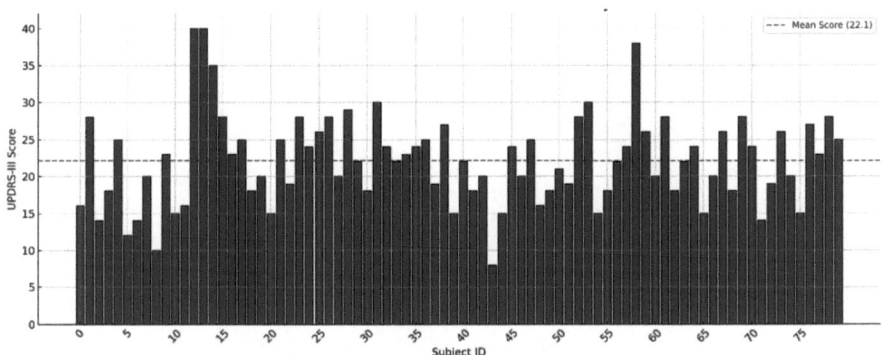

Fig. 3. UPDRS-III Score Distribution Across neuroQWERTY Subjects.

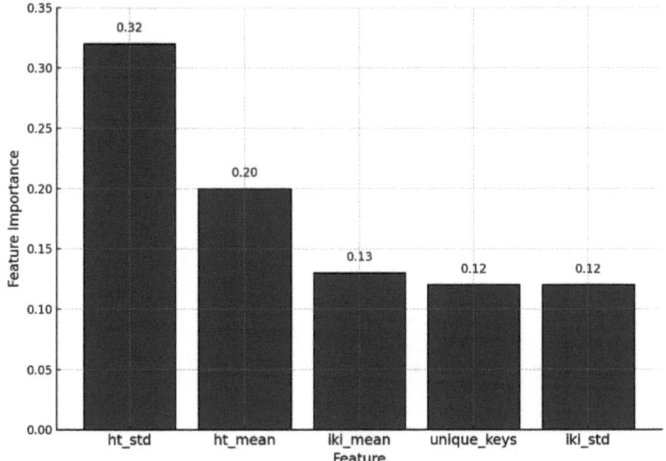

Fig. 4. Feature Importances for PD Classification.

hold time, the more the variability, a phenomenon characteristic of motor delays caused by PD as seen in Fig. 5.

The box plot shows the feature ht_mean, the median of the PD subjects is about 0.16 s with interquartile range of 0.14 and 0.18 s and few outliers stretching to 0.22 s which shows delayed key presses in the motor impaired subjects. In healthy subjects, a median of around 0.12 s, with a smaller IQR of 0.10- 0.14 s and an outlier as far as 0.18 s is observed as seen in Fig. 6.

4.2 Model Performance

The classification activity of three machine learning frameworks; Logistic Regression, Random Forest and XGBoost that came out as a result of training the neuroQWERTY dataset to basically foresee Parkinson's disease (PD) status was tested on a test set of 17 sampling results. Overall performance can be classified as moderate to strong based

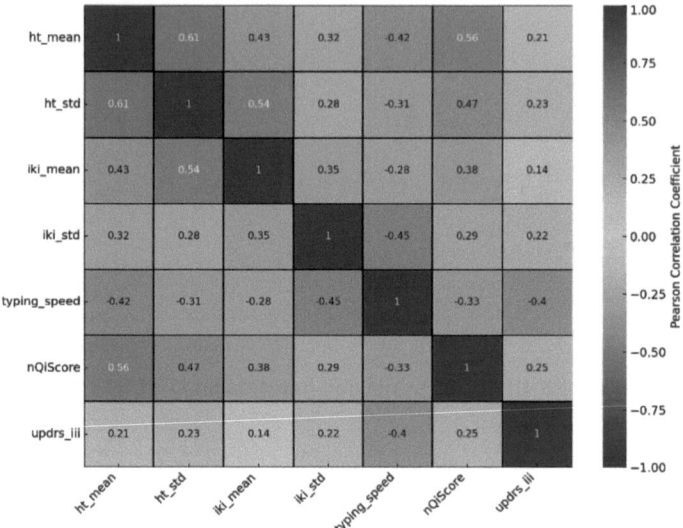

Fig. 5. Feature Importances for PD Classification.

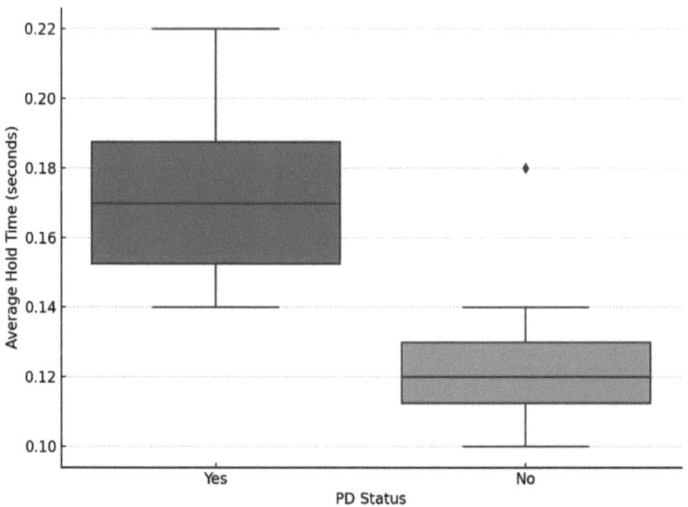

Fig. 6. Box Plot of Hold Time Distribution by PD Status.

on the macro and weighted averages constructing all models, and the Random Forest showed the best performance because of the ability to pick up complex trends in timing features such as ht_std and ht_mean as can be seen in Table 1.

Random Forest model performed well on the ROC curve with AUC 0.96 indicating that it was highly suitable in detecting the PD in real time through keystroke data. XGBoost provided fair AUC of 0.81 which is relatively accurate but less effective implying that it can be optimised. Logistic Regression was the least discriminative one, since its

Table 1. Classification report of the trained models

Model	Accuracy	Precision	Recall	F1-Score
Logistic Regression	0.47	0.50	0.62	0.53
Random Forest	0.88	0.80	0.78	0.89
XGBoost	0.65	0.71	0.75	0.67

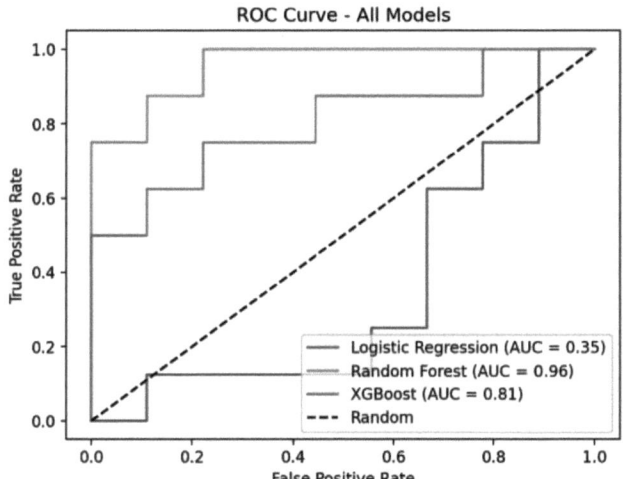

Fig. 7. ROC-AUC curve of the classification Models.

accuracy was 0.47, which evidences the superiority of ensemble methods over Logistic Regression in health monitoring as seen in Fig. 7.

The regression tasks were meant to predict UPDRS-III and nQi scores based on the features (ht_std, ht_mean, iki_mean, iki_std, unique_keys) which were extracted based on the neuroQWERTY dataset. The performance of four regression models—Ridge, LSTM, Random Forest, and ElasticNet—for predicting clinical motor scores from keystroke features. Ridge Regression achieved the best results with the lowest RMSE (0.0304) and highest R^2 (0.7541), indicating strong predictive accuracy and minimal error. LSTM also performed well, highlighting the value of temporal modeling. Random Forest showed moderate performance, while ElasticNet lagged with the lowest R^2 as shown in Table 2.

The most performing model, RidgeCV, in prediction of nQi scores is a scatter graph which consists of actual values of nQi and the predicted values on learning the whole neuroQWERTY data. There is a tight cluster of points along the diagonal line (y = x), which means that there was a high concordance of observed scores and predicted scores with R2 of around 0.65 and RMSE of around 0.15 as seen in Fig. 8.

Table 2. Regression Models performance comparison for

Model	RMSE	R^2 Score
Ridge	0.0304	0.7541
LSTM	0.0312	0.7410
Random Forest	0.0342	0.6888
ElasticNet	0.0422	0.5258

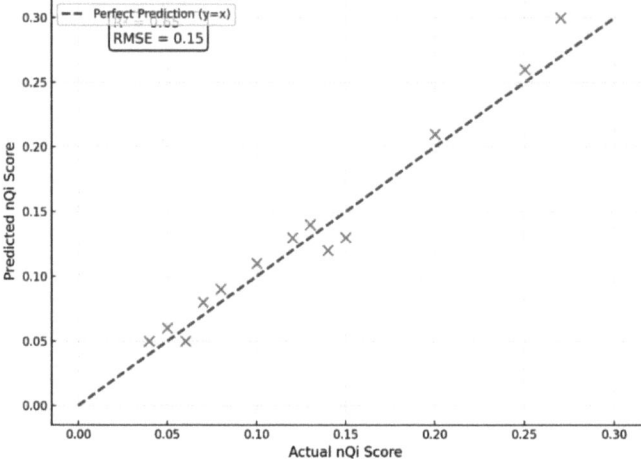

Fig. 8. Actual vs. Predicted Scores for RidgeCV Model

4.3 Live Prediction Interface

The live prediction module also went under intense testing with keystrokes entered by the user being passed to the application through the user-friendly text box interface. The input data were processed dynamically and the main characteristics of the input data ht_mean, ht_std, iki_mean, iki_std, and unique_keys, were extracted, which is used to drive preloaded machine learning models to make predictions immediately, once a sufficient amount of valid keystrokes were received to make further extraction as seen Fig. 9.

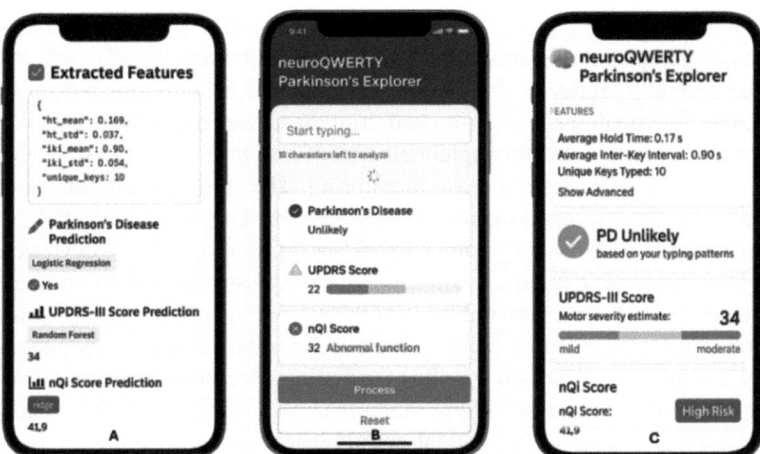

Fig. 9. Live Prediction Module Interface.

The output provided a thorough real-time evaluation in that, a binary model assessed Parkinson's Disease status, an estimated UPDRS-III score that represented the severity of the motor and an nQi, a single value score that provisionally showed us the neuromotor performance with immediate and interpretable data feedback to the user. The models were quite stable in performance as compared to previous tests, and hence confirmed as reliable under varying situations of input. This functional feature strengthens the prospects of the module to be incorporated in practice health monitoring systems where an individual or medical practitioner could use its intuitive user interface to reduce passive and continuous monitoring of the neuromotor changes. The interface has the advantage of being highly responsive and presenting the output results clearly, which makes it highly applicable when used to conduct remote health assessment, especially in low-resource environments or where telehealth is essential, and preludes the development of generalizable, user-centered devices that could assist early identification and ongoing management by performing routine interactions with the device helping in e-health and IoT health monitoring.

5 Conclusion

This study presented the NeuroKey Prediction Tool, a streamlined, web-based diagnostic support system that utilises passive keystroke dynamics to detect and track Parkinson's Disease (PD). The modular Streamlit framework, utilising machine learning models such as Random Forest, Ridge Regression, and LSTM, attained a notable 88% classification accuracy and R^2 scores of 0.75 for predicting clinical motor scores. These findings endorse the practicality of discreet, continuous, and scalable neuromotor monitoring via everyday typing, converting ordinary digital interactions into significant health insights. These findings indicate that passive digital phenotyping using NeuroKey has significant potential to transform Parkinson's care from reactive, clinic-based models to proactive, personalised, and home-centered strategies. Incorporating this technique into pervasive IoT devices can enable early detection and prompt therapies, hence improve quality of life and promoting precision medicine. Stakeholders in e-Health and neurological care should integrate digital biomarkers obtained from routine behaviours, such as typing, into telemedicine workflows and drug management initiatives. Policymakers should promote the research and implementation of non-invasive, accessible technology to provide widespread chronic disease monitoring. The study's limitations encompass a restricted testing setting and insufficient datasets, which may inadequately represent the diversity of real-world typing habits across various demographics, devices, and languages. Future research recommendations include verifying NeuroKey on bigger, more heterogeneous populations, integrating it with real-world mobile typing datasets, and investigating federated learning methodologies to maintain data privacy while enhancing model accuracy. Customised baseline modelling could further improve sensitivity to disease start and progression. These measures will enhance NeuroKey's capacity as a transformational, scalable, and patient-centric instrument for AI-driven neurological treatment.

Disclosure of Interests. The authors have no competing interests to declare that are relevant to the content of this article.

References

1. Aminabee, S.: The future of healthcare and patient-centric care: digital innovations, trends, and predictions. In: Emerging Technologies for Health Literacy and Medical Practice, pp. 240–262. IGI Global Scientific Publishing (2024)
2. Bernardo, J., et al.: EHealth platforms to promote autonomous life and active aging: a scoping review. Int. J. Environ. Res. Public Health **19**(23), 15940 (2022)
3. Javaid, M., Khan, I.H.: Internet of Things (IoT) enabled healthcare helps to take the challenges of COVID-19 Pandemic. J. Oral Biol. Craniofacial Res. **11**(2), 209–214 (2021)
4. Vijayan, V., Connolly, J.P., Condell, J., McKelvey, N., Gardiner, P.: Review of wearable devices and data collection considerations for connected health. Sensors **21**(16), 5589 (2021)
5. Suslu, B., Ali, F., Jennions, I.K.: Understanding the role of sensor optimisation in complex systems. Sensors **23**(18), 7819 (2023)
6. Thakur, R., Saini, A.K., Taliyan, R., Chaturvedi, N.: Neurodegenerative diseases early detection and monitoring system for point-of-care applications. Microchem. J. **208**, 112280 (2025)
7. Dumurgier, J., Tzourio, C.: Epidemiology of neurological diseases in older adults. Revue Neurologique **176**(9), 642–648 (2020)
8. Gupta, R., Srivastava, D., Sahu, M., Tiwari, S., Ambasta, R.K., Kumar, P.: Artificial intelligence to deep learning: machine intelligence approach for drug discovery. Mol. Diversity **25**, 1315–1360 (2021)
9. Giancardo, L., et al.: Computer keyboard interaction as an indicator of early Parkinson's disease. Sci. Rep. **6**(1), 34468 (2016)
10. Karimian-Kelishadrokhi, M., Safi-Esfahani, F.: TD-LSTM: a time distributed and deep-learning-based architecture for classification of motor imagery and execution in EEG signals. Neural Comput. Appl. **36**(25), 15843–15868 (2024)
11. Gheisari, M., et al.: Deep learning: applications, architectures, models, tools, and frameworks: a comprehensive survey. CAAI Trans. Intell. Technol. **8**(3), 581–606 (2023)

12. Khan, A.U., Akram, M., Daniyal, M., Zainab, R.: Awareness and current knowledge of Parkinson's disease: a neurodegenerative disorder. Int. J. Neurosci. **129**(1), 55–93 (2019)
13. Martínez-Martín, P., et al.: Parkinson's disease severity levels and MDS-Unified Parkinson's Disease Rating Scale. Parkinsonism Relat. Disord. **21**(1), 50–54 (2015)
14. McCorkle, R., et al.: Self-management: Enabling and empowering patients living with cancer as a chronic illness. CA: A Cancer J. Clin. **61**(1), 50–62 (2011)
15. Ahmed, M., et al.: End users' and other stakeholders' needs and requirements in the development of a personalized integrated care platform (PROCare4Life) for older people with dementia or Parkinson disease: mixed methods study. JMIR Format. Res. **6**(11), e39199 (2022)
16. Chauhan, R., Singh, D.: Predictive analytics for stress management in nursing: a machine learning approach using wearable IoT devices. In: Singh, D., van 't Klooster, JW., Tiwary, U.S. (eds.) Intelligent Human Computer Interaction. IHCI 2024. LNCS, vol. 15557. Springer, Cham (2025). https://doi.org/10.1007/978-3-031-88705-5_6
17. Chauhan, R., Satyam, A., Yafi, E., Zuhairi, M.F.: Predict the elderly fall using IoT and AI technology. In: International Conference on Cyber Security, Privacy in Communication Networks, pp. 101–113. Singapore: Springer Nature Singapore (2023)
18. Mughal, H., Javed, A.R., Rizwan, M., Almadhor, A.S., Kryvinska, N.: Parkinson's disease management via wearable sensors: a systematic review. IEEE Access **10**, 35219–35237 (2022)
19. Chauhan, R., Avasthi, S., Alankar, B., Kaur, H.: Smart IoT systems: data analytics, secure smart home, and challenges. In: Transforming the Internet of Things for Next-Generation Smart Systems, pp. 100–119. IGI Global (2021)
20. Chauhan, R., Gogna, D., Avasthi, S.: Deep learning approach for the prediction of skin diseases. In: Machine Learning Models and Architectures for Biomedical Signal Processing, pp. 301–318. Academic Press (2025)
21. Rashidi, H.H., Tran, N., Albahra, S., Dang, L.T.: Machine learning in health care and laboratory medicine: general overview of supervised learning and Auto-ML. Int. J. Lab. Hematol. **43**, 15–22 (2021)
22. NeuroQWERTY MIT-CSXPD Dataset V1.0.0 (2016). https://physionet.org/content/nqmitcsxpd/1.0.0/
23. Alfalahi, H., et al.: Diagnostic accuracy of keystroke dynamics as digital biomarkers for fine motor decline in neuropsychiatric disorders: a systematic review and meta-analysis. Sci. Rep. **12**(1), 7690 (2022)

Open Access This chapter is licensed under the terms of the Creative Commons Attribution 4.0 International License (http://creativecommons.org/licenses/by/4.0/), which permits use, sharing, adaptation, distribution and reproduction in any medium or format, as long as you give appropriate credit to the original author(s) and the source, provide a link to the Creative Commons license and indicate if changes were made.

The images or other third party material in this chapter are included in the chapter's Creative Commons license, unless indicated otherwise in a credit line to the material. If material is not included in the chapter's Creative Commons license and your intended use is not permitted by statutory regulation or exceeds the permitted use, you will need to obtain permission directly from the copyright holder.

Building Trust: Privacy, Security and Responsible AI

Enhancing Security and Privacy in Federated Learning for Distributed Systems: The REMINDER Approach

Francisco J. Cortés-Delgado[1(✉)], Enrique Mármol Campos[1], José L. Hernández-Ramos[1], Antonio Skarmeta[1], Shahid Latif[2], Djamel Djenouri[2], Stephan Krenn[3], Andrei Puiu[4], and Anamaria Vizitiu[4]

[1] University of Murcia, Murcia, Spain
{franciscojose.cortesd,enrique.marmol,jluis.hernandez,skarmeta}@um.es
[2] University of the West of England, Bristol, UK
{Shahid.Latif,Djamel.Djenouri}@uwe.ac.uk
[3] AIT Austrian Institute of Technology, Vienna, Austria
stephan.krenn@ait.ac.at
[4] Siemens SRL, Foundational Technologies, Transilvania University of Brasov, Brasov, Romania
{andrei.puiu,anamaria.vizitiu}@siemens.com

Abstract. Federated Learning (FL) enables collaborative model training without centralizing raw data, but distributed deployments remain exposed to typical poisoning and inference attacks and must operate across resource-constrained edge environments. The REMINDER project addresses these challenges by designing an edge-centric framework that provides privacy and security mechanisms with byzantine-robust learning approaches. This paper reports some of the project's mechanisms and their implications for the development of robust and secure FL deployments, including: (i) a threat model addressing poisoning and inference risks; (ii) a modular architecture with differential privacy, secure authenticated updates, and robust aggregation against malicious clients; and (iii) two representative validation scenarios, such as eHealth and smart buildings, which ground design choices and highlight domain-specific constraints. Building on these contributions, the present work formalizes the end-to-end workflow, specifies component interfaces, and links attack classes to concrete mitigations within REMINDER, while outlining open challenges such as verifiable aggregation and the privacy–utility trade-off introduced by differential privacy in common FL settings.

Keywords: Artificial Intelligence · Federated Learning · Data Privacy · Distributed Systems · Edge Computing · Smart Buildings · Healthcare

1 Introduction

Artificial Intelligence (AI) is a key driver of technological innovation, enabling real-time data-driven decision-making across multiple domains. Its role is critical in achieving the UN's Sustainable Development Goals (SDGs) for 2030 [30]. The rapid progress of AI is mainly supported by hyperconnectivity, which facilitates the seamless collection and exchange of data from multiple sources. In this context, the integration of Internet of Things (IoT) technologies and 5G/6G networks integrating distributed systems into the Internet infrastructure foster the development of smart cities [33] and advanced eHealth services [12].

IoT-based distributed systems interconnect physical devices to enable real-time data exchange, empowering industries to make informed decisions [24]. However, these systems often rely on centralized AI architectures that require data sharing with central servers, raising significant privacy concerns. FL [22] offers a promising solution by allowing collaborative model training directly on local data. Each participant trains a model on private datasets and shares only model parameters, preserving privacy. This approach enables organizations to develop generalizable models while mitigating data exposure risks. A global model is aggregated from local updates and redistributed, assuming a trusted aggregation server.

However, due to the decentralized nature of FL, it has to face several challenges. On the one hand, FL remains vulnerable to security threats, such as poisoning attacks [2], where adversaries inject malicious updates to degrade model performance through aggregation. On the other hand, FL is exposed to privacy risks such as inference attacks, which can expose sensitive data through the training weights during the communication between clients and the server. Addressing these challenges requires robust aggregation methods, and privacy-preserving techniques, including Differential Privacy (DP) and cryptographic protocols [23].

REMINDER is a CHIST-ERA transnational project (Call 2022, SPiDDS)[1] with partners in Spain, Austria, Romania, and the UK. Its objective is to provide an edge-centric FL framework that integrates Differential Privacy (DP), Secure Multi-Party Computation (SMPC), and robust aggregation. This paper explores the project's architecture, components, and use cases in eHealth and smart buildings.

Paper Outline. The remainder of this paper is organized as follows: Sect. 2 outlines FL properties and challenges. Section 3 discusses potential attacks and REMINDER's mitigation strategies. Section 4 details the framework architecture. Section 5 presents use cases, and Sect. 6 concludes with future research directions.

---

[1] https://www.chistera.eu/projects/reminder

2 Properties and Challenges of FL

FL is a transformative approach to ML that enhances data privacy by keeping training data on local devices, minimizing exposure risks [22]. FL settings typically comprise two different entities: *clients* and *aggregators*. Each client possesses unique, private datasets undisclosed to others during the collaborative training process. In FL, training can be performed on compute-capable surrogate devices that did not originally collect the data, or on edge devices as part of an edge-computing deployment [27]. By processing data locally, edge computing provides the on-device computation and networking needed for resource-efficient learning: it reduces end-to-end latency and backhaul bandwidth, keeps sensitive data in situ to strengthen privacy, improves resilience to intermittent connectivity, and enables faster, context-aware personalization. After this training process, they solely exchange the resulting weights with the aggregator. Next, the aggregator, or server, aggregates these weights and redistributes them for subsequent training. To combine or aggregate such weights, the server uses an *aggregation function*, whose result will be provided to the clients. For instance, the most common aggregation function called FedAvg consists of simply taking the average of the clients' weights [22].

However, despite the advances in FL [21], FL still has to face several security and privacy challenges. First, the clients' training phase occurs locally, so the server does not have direct access to the raw data or the exact training procedure. Although the server does receive model updates from the clients, it lacks the contextual information needed to verify their correctness or authenticity. This makes it difficult to detect malicious behavior from compromised (*Byzantine*) clients, whose goal is to degrade performance or delay model convergence through poisoning attacks [2]. Poisoning attacks deliberately change either the dataset of a client *(data poisoning)* or the weights sent to the server *(model poisoning)*. Indeed, several works have proved the vulnerability of FL to poisoning attacks [1,3], especially when FedAvg is used as the aggregation function.

Data poisoning can be classified in two ways. The first classification is based on how the malicious data is injected [26]: either through a separate client with malicious data *(Sybil)* or by directly modifying a benign client's data *(Backdoor)*. The second classification considers whether the labels of the client datasets are altered [29]. Clean-label attacks maintain the original labels while modifying other features, whereas dirty-label attacks manipulate both the features and the labels. Regarding model poisoning attacks, they are generally also divided into two categories: untargeted and targeted poisoning attacks [34]. *Untargeted attacks* introduce minimal modifications to the global model to cause incorrect predictions broadly, while *targeted attacks* are designed to manipulate predictions on specific inputs while maintaining the model's accuracy on benign data. Untargeted poisoning attacks can be further categorized based on whether malicious clients act independently or collude [28]. In collusion-based attacks, malicious clients first proceed with standard model training independently. However, after training, they combine their weight updates by averaging them and adding a penalty function to harm the aggregated model.

Moreover, although FL is mainly proposed as a privacy-preserving approach, it still faces significant privacy concerns. One major issue is that clients' weights can be intercepted during communication between the server and the clients, and with inference attacks it is possible to reconstruct the dataset used. Indeed, in a typical FL setting, the server is usually able to access the weights uploaded by the clients throughout the rounds. Inference attacks are classified as model inversion, membership attacks, and reconstruction attacks depending on the target information and techniques employed [26]. In *model inversion* attacks, an adversary leverages the model's architecture and parameters to reconstruct original inputs from outputs or gradients, posing risks in over-parameterized or overfitted models that may memorize sensitive data. *Membership inference attacks* determine whether a specific data point was part of the training set by analyzing confidence scores, loss values, or gradient norms, thereby revealing private participation in the model's training. In contrast, *reconstruction attacks* use gradient matching techniques to recover entire datasets or their statistical properties from shared gradients, especially in federated learning environments. Together, these categories highlight the diverse vulnerabilities in machine learning systems and the critical need for robust privacy-preserving measures.

3 Mitigating Attacks in FL Deployments

As explained before, FL is vulnerable to security and privacy threats that can compromise model integrity and confidentiality. We next explain the different countermeasures for these attacks, and which of them REMINDER applies.

3.1 Enhancing the Security in FL

There are two main approaches to defend against poisoning attacks: applying robust aggregation functions, and implementing clustering techniques to identify malicious clients. On one hand, robust aggregation functions reduce the impact of outlier values. Examples of these functions are the median, the trimmed mean, and more sophisticated data transformation techniques, such as the Fast Fourier Transform (FFT), which was used in our previous work [5]. Other works such as FoolsGold [11] or FLTrust [7] implement techniques to redefine the weighting of clients in the aggregation process in order to reduce the influence of malicious clients. On the other hand, clustering-based approaches aim to explicitly identify and remove malicious clients before applying FedAvg. Many existing methods, including SignGuard [35] and FedDMC [25], use techniques that increase the dissimilarity between benign and malicious clients to separate them into two clusters. The larger cluster is then assumed to contain the benign clients.

Based on the previous description, REMINDER focuses on mitigating Sybil attacks, as well as one dirty-label attack (label flipping), since these attacks tend to be more impactful than clean-label ones. In addition, REMINDER addresses five types of untargeted model poisoning attacks, including random, LIE, STATOPT, min-max, and min-sum [28], which are broadly studied and have proven highly effective. Untargeted attacks are particularly relevant because prior research [4] treats data poisoning as a specific case of local model poisoning, and these attacks tend to be more difficult to detect than targeted ones [19].

3.2 Enhancing Privacy in FL

To safeguard model weights against inference attacks, multiple privacy-preserving techniques are employed to mitigate various threats. Differential Privacy (DP) [18] can be employed to obfuscate either the training data or model updates, giving statistical privacy which guarantees over the data against an adversary. In particular, DP adds noise following different distributions in the data to increase the difficulty of inferring data. Next, Secure Multi-party Computation (SMPC) [10] allows multiple participants to jointly compute a function over their private inputs while keeping these inputs confidential. The computation is decomposed into operations on cryptographic shares, meaning each participant holds only a fragment of the data. This design ensures that no single party can reconstruct the entire input, and only the final result is disclosed, thereby maintaining the privacy of the individual inputs. Finally, Homomorphic Encryption (HE) [8] is a cryptographic technique that allows computations to be performed directly on encrypted data. This means that the results of these operations, once decrypted, match those obtained by performing the operations on the original plaintexts, thereby preserving data confidentiality throughout the computation process.

REMINDER will analyze different types of DP mechanisms to mitigate membership and reconstruction attacks. To address model inversion attacks, we will consider several encryption techniques to enhance privacy protection.

4 REMINDER Architecture

The REMINDER methodology presents a privacy-preserving and secure FL framework to overcome the main threats in federated scenarios. Our proposed architecture is presented in Fig. 1. In this case, during the communication between clients and server we add privacy-preserving techniques such as DP, encryption methods and signature processes. And then, the server applies different aggregation functions and frameworks for detecting malicious clients. In particular, the procedure of REMINDER will be:

(At clients' side)

1. We train the model in the "Trainer", a device which is technically capable of this task
2. After the training, the clients will apply DP techniques to obfuscate the weights.
3. The clients will encrypt to add a higher degree of privacy. These updates are encrypted using a cipher. This ensures that, even if intercepted by a malicious entity, it would be significantly more challenging to infer any sensitive information.

4. The model updates are signed to ensure their integrity and authenticity. The model updates are then sent to the server,

(At servers' side)

5. The server performs a malicious client detector to make a first filter in order to remove as many malicious clients as possible.
6. The server aggregates all the updates received from the clients using a robust aggregation function, which discards malicious updates. The aggregation should be performed on encrypted data, ensuring that the server never has access to the actual weight values.

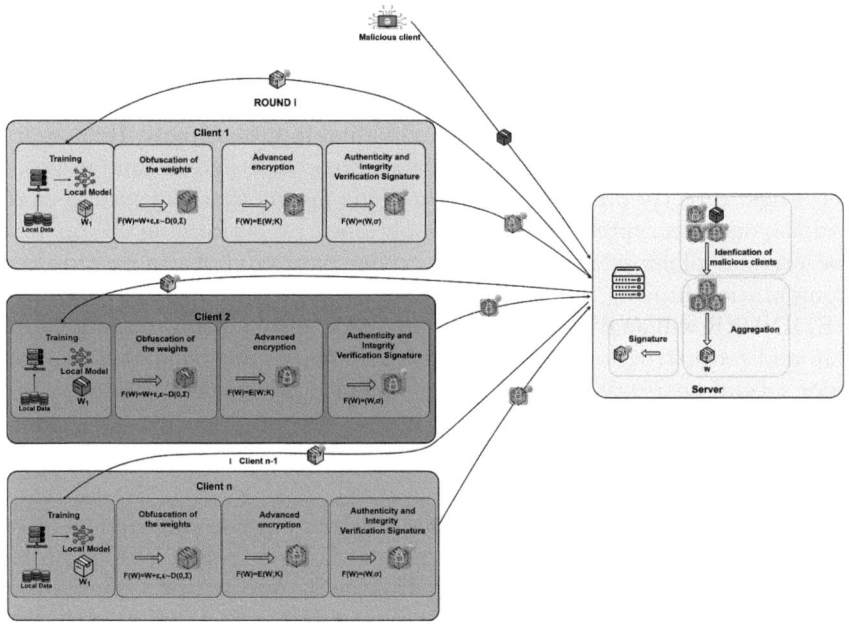

Fig. 1. REMINDER Architecture

4.1 Clients' Modules

In REMINDER framework, we suppose that a client is a "trainer", a device which is technically capable of training an ML model, where the data collection can be performed by another device. After the training of the clients, several modules are applied to protect the weights from any inference attack. First, the weights are obfuscated using DP techniques, since DP is often preferred in FL settings due to the high communication or computation demands of, e.g., SMPC or FHE. Next, the clients will encrypt its data before sending it to the server, to ensure that, even if intercepted by a malicious entity, it becomes computationally infeasible to infer sensitive information. Finally, to assure integrity and authenticity in the system, the client digitally signs its message, e.g., using an Elliptic Curve Cryptography (ECC) based scheme [9], to ensure that fake updates produced by Sybil attacks do not affect the aggregation. In particular, ECC helps contain poisoning attacks in FL since that guarantees who sends updates and what was actually sent. Clients sign their model updates with ECC and the server verifies signatures to block forged updates generated by, for instance, Sybil clients.

Conversely, in our threat model, Byzantine clients pursue indiscriminate degradation of the global model by either corrupting their local data or submitting poisoned updates. These attacks operate online, i.e., they are executed in any training round, and adversaries may collude by sharing their local updates to amplify impact. Adversaries possess white-box knowledge of the models produced by themselves (and any colluding malicious peers) each round, but cannot observe benign clients' updates and are assumed unaware of the server's aggregation rule. Their capabilities include manipulating local datasets (data poisoning) or altering gradients/weights before submission (model poisoning), achievable via device compromise or in-transit interception.

4.2 Server's Strategy

Upon receiving messages from clients in a specific training round, the server validates the signatures and discards messages otherwise; furthermore, it decrypts the messages. In order to mitigate the impact of the possible malicious clients in the system, the server applies a malicious clients detector and then, a robust aggregation function. Indeed, although robust aggregation functions protect systems against Byzantine clients, they have some limitations [7]. For instance, their performance tends to decline as the number of malicious clients increases, or it is needed to specify the number of malicious clients in the system to leverage the maximum performance. Likewise, in malicious client detectors, as [35] suggests, there can be rounds when some malicious clients manage to overcome the latter identifying process. Therefore, malicious clients may harm the global model despite the previous efforts. In this direction, we first apply a malicious client detector to reduce the number of malicious clients as much as possible. Next, we apply a robust aggregation function to mitigate the effect of the possible remaining malicious clients. As the total number of malicious clients is reduced, that mitigates the drawback of robust aggregation functions. Before sending the updated model parameters back to the clients, the server again signs the result, such that the clients have guarantees that those are indeed legitimate.

From a privacy perspective, in the initial architecture, only limited trust in the server is necessary due to the usage of DP. However, regarding authenticity, the server needs to be fully trusted in the sense that the clients do not have any guarantees on the validity of the received model parameters. During the evolution of the REMINDER project, we will thus also investigate means to ensure authenticity and integrity by expanding the involved signature modules. One option could be that the server's signature module not only signs the updated model parameters, but also proves that they were indeed derived from individual updates received from, and signed by, clients in the system. This could be achieved by computing a corresponding non-interactive zero-knowledge proof of knowledge (NIZK) [13] or a zero-knowledge succinct non-interactive argument of knowledge (zkSNARK) [15]. While this is efficiently doable for simple aggregation functions such as FedAvg, it remains an open research challenge to also cover complex malicious clients detectors and more robust aggregation functions.

5 REMINDER Use Cases

The REMINDER framework aims to address critical security and privacy challenges in FL through innovative solutions tailored for real-world applications. We next present two representative use cases around eHealth and smart buildings, chosen to demonstrate the practicality and effectiveness of the proposed framework. These scenarios highlight how REMINDER leverages its architecture and privacy-preserving techniques to meet specific requirements in diverse and sensitive domains.

5.1 eHealth

Healthcare data are often distributed across multiple sites, especially in hospitals, due to the complexity of healthcare systems and processes. Strict regulations like HIPAA[2] and GDPR[3] protect patient privacy, yet however limit data sharing, thereby hindering the development of advanced ML technologies, which rely on large datasets. Additionally, running ML on vast data requires significant computation, challenging resource-limited environments. FL addresses these issues by allowing Electronic Health Record (EHR) data from medical institutions to collaboratively train ML models without sharing raw data, enhancing privacy.

Researchers have made significant efforts to enhance security in FL for healthcare [32], reduce communication costs [37], lower computation requirements [31], and improve model performance [20]. However, current solutions remain in early stages and lack validation in real-world settings. The architecture of REMINDER is particularly suited for the healthcare domain. Healthcare providers can deploy REMINDER nodes at the edge of their networks, where sensitive EHRs are stored.

To showcase REMINDER's potential in tackling the challenges of developing AI-driven healthcare solutions, we explore two specific use cases over the course of the project.

---

[2] https://www.hhs.gov/hipaa/index.html.
[3] https://gdpr-info.eu/.

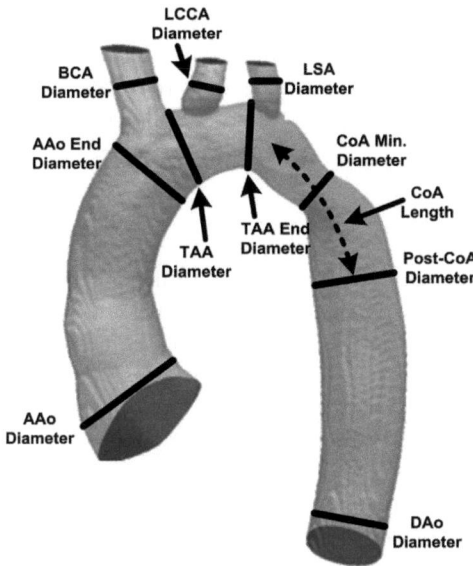

Fig. 2. Reduced set of anatomical features that can be leveraged by a DL algorithm.

Atrial Fibrillation. Wearable medical devices, powered by data analytics, cloud computing, and machine learning (ML), have become vital tools for early disease detection. A key application is heart monitoring, where these devices help identify atrial fibrillation—a prevalent cardiac arrhythmia linked to irregular and rapid heartbeats, often leading to strokes. To detect such conditions, wearable sensors capture data like electrocardiograms (ECGs), which are then transmitted to external service providers for analysis using ML algorithms. However, the collection, sharing, and processing of sensitive health data raise critical privacy and security concerns, as any breach could compromise patient confidentiality.

Aortic Coarctation. Aortic coarctation is a congenital heart defect characterized by a narrowing of the aortic media, reducing the aortic lumen. It accounts for 5 to 8% of all congenital heart disease cases. Diagnosis typically relies on assessing the pressure gradient across the coarctation, which can be estimated using cuff-derived pressure differences between upper and lower extremities, Doppler-based measurements via the Bernoulli equation, or catheter-based measurements—the latter being the gold standard.

To offer a non-invasive alternative to catheterization, [17] introduced personalized blood flow simulations using computational fluid dynamics (CFD) to estimate the pressure drop. However, this method has notable drawbacks: (1) anatomical model reconstruction from medical images requires semi-automated steps with manual adjustments, and (2) CFD simulations are computationally expensive and time-consuming.

A promising solution is to use a reduced set of fully automated anatomical measurements (see Fig. 2) and leverage deep learning models to predict CFD outputs. Nevertheless, training these models requires access to patient-derived anatomical data, which is subject to strict regulatory controls designed to safeguard patient privacy.

5.2 Smart Buildings

The integration of IoT devices into smart buildings has revolutionized energy consumption management by enabling real-time environmental control. These systems monitor temperature, humidity, and occupancy to optimize heating, ventilation, and air conditioning (HVAC) operations, ensuring comfort while minimizing energy waste. However, privacy concerns arise when energy usage patterns expose sensitive information about occupants' behaviors, such as daily routines or room occupancy [14]. FL emerges as a viable solution, allowing buildings to collaboratively train energy optimization models without sharing raw data, thereby reducing privacy risks while improving predictive accuracy [6].

A significant challenge in FL deployment across buildings is the heterogeneity of sensor infrastructures. Some buildings are well-equipped with advanced IoT devices tracking detailed energy consumption, while others lack comprehensive instrumentation. To address this, Federated Transfer Learning (FTL) facilitates knowledge transfer between highly-instrumented buildings and those with limited data availability, improving model generalization [6].

The Pleiades Building at the University of Murcia serves as a real-world testbed, providing extensive datasets on HVAC usage, indoor temperatures, weather conditions, and occupant movement [16]. By clustering buildings with similar energy profiles, FL models can be trained efficiently on well-instrumented sites and adapted to new environments through domain adaptation techniques [36]. This multi-source knowledge fusion enhances model robustness and applicability.

Pleiades' Architecture. The architecture can be seen in Fig. 3. The Pleiades Building Management System (BMS) manages HVAC units, temperature and humidity sensors, and lighting systems, transmitting data in JSON format. This system enables real-time dynamic energy optimization using smart meters. This IoT-enabled infrastructure underscores the potential of FL for privacy-preserving energy optimization. Data collected from multiple buildings serves as a benchmark for training federated models that support adaptive and scalable energy management strategies.

Risks and Mitigation Strategies. While FL preserves privacy by eliminating centralized data storage, risks persist in analyzing energy usage patterns. Adversaries could infer occupant behavior by correlating energy consumption anomalies with external data sources such as CCTV footage, WiFi logs, or social media check-ins [14]. For example, a sudden drop in energy usage after

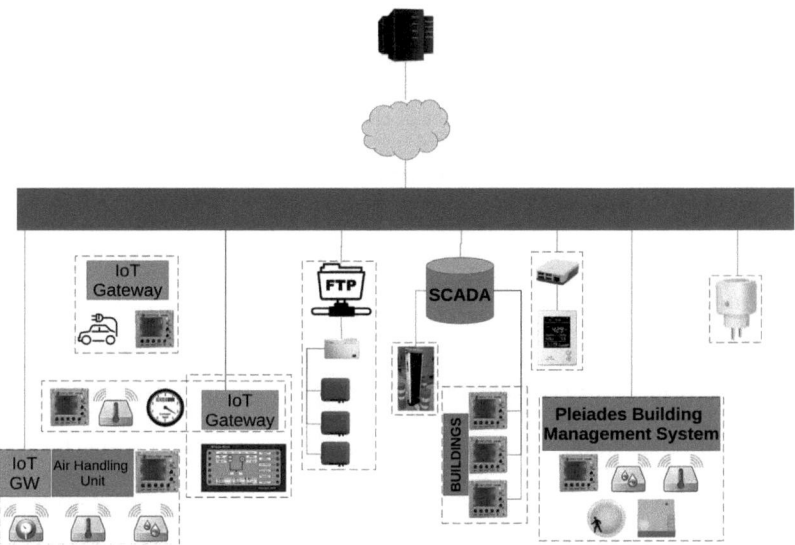

Fig. 3. Pleiades architecture diagram

an individual posts "leaving work" could reveal movement patterns. To counteract these threats, REMINDER integrates multiple security layers, including authentication mechanisms and techniques to identify and mitigate the impact of malicious or compromised clients. Furthermore, DP adds controlled noise to gradient updates, ensuring individual energy profiles remain obfuscated.

This comprehensive approach ensures that REMINDER supports secure, efficient, and privacy-preserving FL for smart buildings, facilitating large-scale, intelligent energy management while safeguarding occupant privacy.

6 Conclusion and Future Directions

The REMINDER approach introduces a privacy-preserving framework for FL in decentralized systems. By leveraging cryptographic techniques like DP, as well as advanced encryption and signature schemes, along with an edge-based architecture and robust aggregation, it tackles key security and privacy challenges in collaborative ML. This enables sectors like eHealth and smart buildings to use FL without compromising data or violating regulations like GDPR. Innovative client authentication and dynamic node selection ensure only legitimate systems contribute to training, enhancing security in heterogeneous, resource-limited environments. At this stage, the project primarily defines the architecture and foundational mechanisms. Over the course of the project, these conceptual elements will be progressively integrated, tested, and refined in real-world scenarios. This phased evolution seeks to ensure that the proposed solutions remain adaptable and effective for different industrial and societal needs. Ultimately, REMINDER aims to offer a scalable, trustworthy FL ecosystem capable of seamless deployment across multiple domains, potentially extending beyond eHealth and smart buildings to sectors like finance and autonomous systems.

Acknowledgements. This work is part of the REMINDER project, funded under the EU CHIST-ERA initiative (Grant EP/Y036301/1 from EPSRC, UK, PCI2023-145989-2 funded by MICIU/AEI/10.13039/501100011033, Austrian Science Fund (FWF): I 6650-N, and by the European Union NextGenerationEU/PRTR).

References

1. Bagdasaryan, E., Veit, A., Hua, Y., Estrin, D., Shmatikov, V.: How to backdoor federated learning. In: International Conference on Artificial Intelligence and Statistics, pp. 2938–2948. PMLR (2020)
2. Barreno, M., Nelson, B., Sears, R., Joseph, A.D., Tygar, J.D.: Can machine learning be secure? In: AsiaCCS 2006, pp. 16–25 (2006)
3. Baruch, G., Baruch, M., Goldberg, Y.: A little is enough: circumventing defenses for distributed learning. In: Advances in Neural Information Processing Systems, vol. 32 (2019)
4. Bhagoji, A.N., Chakraborty, S., Mittal, P., Calo, S.: Analyzing federated learning through an adversarial lens. In: International Conference on Machine Learning, pp. 634–643. PMLR (2019)
5. Campos, E.M., Vidal, A.G., Ramos, J.L.H., Skarmeta, A.: FedRDF: a robust and dynamic aggregation function against poisoning attacks in federated learning. arXiv preprint arXiv:2402.10082 (2024)
6. Campos, E.M., Vidal, A.G., Hernández Ramos, J.L., Skarmeta, A.: federated transfer learning for energy efficiency in smart buildings. In: IEEE INFOCOM 2023, pp. 1–6 (2023). https://doi.org/10.1109/INFOCOMWKSHPS57453.2023.10225844
7. Cao, X., Fang, M., Liu, J., Gong, N.Z.: FLtrust: byzantine-robust federated learning via trust bootstrapping. arXiv preprint arXiv:2012.13995 (2020)
8. Chang, W., Zhu, T.: Gradient-based defense methods for data leakage in vertical federated learning. Comput. Secur. **139**, 103744 (2024)
9. Chen, Y., Su, Y., Zhang, M., Chai, H., Wei, Y., Yu, S.: FEDTOR: an anonymous framework of federated learning in internet of things. IEEE Internet Things J. **9**(19), 18620–18631 (2022). https://doi.org/10.1109/JIOT.2022.3162826
10. Choudhury, O., et al.: Anonymizing data for privacy-preserving federated learning. arXiv preprint arXiv:2002.09096 (2020)
11. Fung, C., Yoon, C.J., Beschastnikh, I.: Mitigating sybils in federated learning poisoning. arXiv preprint arXiv:1808.04866 (2018)
12. Gao, X., He, P., Zhou, Y., Qin, X.: Artificial intelligence applications in smart healthcare: a survey. Future Internet **16**(9) (2024). https://doi.org/10.3390/fi16090308, https://www.mdpi.com/1999-5903/16/9/308
13. Goldwasser, S., Micali, S., Rackoff, C.: The knowledge complexity of interactive proof-systems (extended abstract). In: Sedgewick, R. (ed.) STOC, pp. 291–304. ACM (1985). https://doi.org/10.1145/22145.22178
14. González-Vidal, A., Ramallo-González, A.P., Skarmeta, A.: Empirical study of massive set-point behavioral data: towards a cloud-based artificial intelligence that democratizes thermostats. In: IEEE SMARTCOMP 2018, pp. 211–218 (2018). https://doi.org/10.1109/SMARTCOMP.2018.00093

15. Groth, J.: On the size of pairing-based non-interactive arguments. In: Fischlin, M., Coron, J.-S. (eds.) EUROCRYPT 2016. LNCS, vol. 9666, pp. 305–326. Springer, Heidelberg (2016). https://doi.org/10.1007/978-3-662-49896-5_11
16. Ibarra, A.M., González-Vidal, A., Skarmeta, A.: PLEIAData: consumption, HVAC, temperature, weather and motion sensor data for smart buildings applications. Sci. Data **10**, 118 (2023). https://doi.org/10.1038/s41597-023-02023-3
17. Itu, L., et al.: Non-invasive hemodynamic assessment of aortic coarctation: validation with in vivo measurements. Ann. Biomed. Eng. **41**(4), 669–681 (2013). https://doi.org/10.1007/s10439-012-0715-0
18. Ji, Z., Lipton, Z.C., Elkan, C.: Differential privacy and machine learning: a survey and review. arXiv preprint arXiv:1412.7584 (2014)
19. Kiourti, P., Wardega, K., Jha, S., Li, W.: TrojDRL: evaluation of backdoor attacks on deep reinforcement learning. In: ACM/IEEE DAC, pp. 1–6. IEEE (2020)
20. Li, T., Sahu, A.K., Talwalkar, A., Smith, V.: Federated learning: challenges, methods, and future directions. IEEE Signal Process. Mag. **37**(3), 50–60 (2020). https://doi.org/10.1109/MSP.2020.2975749
21. Kwon, S., Choi, J.: An agent-based adaptive monitoring system. In: Shi, Z.-Z., Sadananda, R. (eds.) PRIMA 2006. LNCS (LNAI), vol. 4088, pp. 672–677. Springer, Heidelberg (2006). https://doi.org/10.1007/11802372_77
22. McMahan, B., Moore, E., Ramage, D., Hampson, S., y Arcas, B.A.: Communication-efficient learning of deep networks from decentralized data. In: Artificial intelligence and statistics, pp. 1273–1282. PMLR (2017)
23. Mothukuri, V., Parizi, R.M., Pouriyeh, S., Huang, Y., Dehghantanha, A., Srivastava, G.: A survey on security and privacy of federated learning. Futur. Gener. Comput. Syst. **115**, 619–640 (2021)
24. Mu, X., Antwi-Afari, M.F.: The applications of internet of things (IoT) in industrial management: a science mapping review. Int. J. Prod. Res. **62**(5), 1928–1952 (2024). https://doi.org/10.1080/00207543.2023.2290229
25. Mu, X., et al.: FEDDMC: efficient and robust federated learning via detecting malicious clients. IEEE TDSC (2024)
26. Neto, H.N.C., Hribar, J., Dusparic, I., Mattos, D.M.F., Fernandes, N.C.: A survey on securing federated learning: analysis of applications, attacks, challenges, and trends. IEEE Access **11**, 41928–41953 (2023)
27. Ranjan, R., Buyya, R., Dehghantanha, A., Mollah, M.B.A., Yu, J., Ghosh, U.: Edge computing and federated learning for privacy-preserving healthcare system: state-of-the-art and future perspectives. Smart Health **20** (2021). https://doi.org/10.1016/j.smhl.2021.100200, https://www.sciencedirect.com/science/article/pii/S266729522100009X
28. Shejwalkar, V., Houmansadr, A.: Manipulating the byzantine: Optimizing model poisoning attacks and defenses for federated learning. In: NDSS (2021)
29. Tian, Z., Cui, L., Liang, J., Yu, S.: A comprehensive survey on poisoning attacks and countermeasures in machine learning. ACM Comput. Surv. **55**(8), 1–35 (2022)
30. Vinuesa, R., et al.: The role of artificial intelligence in achieving the sustainable development goals. Nat. Commun. **11**(1), 1–10 (2020)
31. Wang, H., Yurochkin, M., Sun, Y., Papailiopoulos, D., Khazaeni, Y.: Federated learning with matched averaging (2020). https://arxiv.org/abs/2002.06440
32. Wei, K., et al.: Federated learning with differential privacy: algorithms and performance analysis. IEEE Trans. Inf. Forensics Secur. **15**, 3454–3469 (2020)
33. Wolniak, R., Stecuła, K.: Artificial intelligence in smart cities–applications, barriers, and future directions: a review. Smart Cities **7**(3), 1346–1389 (2024)

34. Xia, G., Chen, J., Yu, C., Ma, J.: Poisoning attacks in federated learning: a survey. IEEE Access **11**, 10708–10722 (2023)
35. Xu, J., Huang, S.L., Song, L., Lan, T.: Byzantine-robust federated learning through collaborative malicious gradient filtering. In: IEEE ICDCS 2022, pp. 1223–1235. IEEE (2022)
36. Zafeiropoulos, A., Fotopoulou, E., González-Vidal, A., Skarmeta, A.: Detaching the design, development and execution of big data analysis processes: a case study based on energy and behavioral analytics. In: GIoTS 2018, pp. 1–6 (2018). https://doi.org/10.1109/GIOTS.2018.8534525
37. Zhang, W., et al.: Dynamic-fusion-based federated learning for Covid-19 detection. IEEE Internet Things J. **8**(21), 15884–15891 (2021). https://doi.org/10.1109/JIOT.2021.3056185

Open Access This chapter is licensed under the terms of the Creative Commons Attribution 4.0 International License (http://creativecommons.org/licenses/by/4.0/), which permits use, sharing, adaptation, distribution and reproduction in any medium or format, as long as you give appropriate credit to the original author(s) and the source, provide a link to the Creative Commons license and indicate if changes were made.

The images or other third party material in this chapter are included in the chapter's Creative Commons license, unless indicated otherwise in a credit line to the material. If material is not included in the chapter's Creative Commons license and your intended use is not permitted by statutory regulation or exceeds the permitted use, you will need to obtain permission directly from the copyright holder.

Towards a Responsible AI Adoption/Adaptation (RAA) Ecosystem: Vision and Model to Keep Socio-Technological Balance

Parwinder Singh[✉][iD], Asim Ul Haq[iD], and Mirko Presser[iD]

Department of Business Development and Technology, Aarhus University, 7400 Herning, Denmark
{parwinder,asimulhaq,mirko}@btech.au.dk

Abstract. The rapid adoption of AI technologies is outpacing our ability to assess their long-term societal and economic impacts. Initially, AI was expected to automate only repetitive, low-skill tasks. However, presently, highly skilled roles such as in software development, manufacturing, and finance are being automated. Tasks that once required many professionals can now be managed by a few, disproportionately benefiting those with greater resources and economic power, such as large corporations. This trend may lead to a growing socio-technological imbalance, for instance, a growing mismatch between the rapid advancements in artificial intelligence and society's ability to adapt to and govern these changes in a fair, ethical, and inclusive manner. As AI-driven automation is increasingly being adopted across almost all business domains, covering all process-level functions. Additionally, there is a lack of practical methods for the ethical and responsible adoption of AI, which are either not being implemented or not well understood by stakeholders in the business ecosystem. This emphasizes the role of Responsible AI (RAI), which is becoming critical and essential in addressing the socio-technological imbalance in the organizational context regarding adopting/adapting AI. RAI ensures that systems are developed and deployed with core principles such as explainability, fairness, ethics, transparency, accountability, human oversight, and privacy. Furthermore, AI itself presents as a technological capability to promote responsible behavior by assessing, predicting, supporting, and regulating its own societal (i.e., organization-specific) and technological impacts. This study explores this dual perspective by reviewing the literature to propose a RAI Adoption/Adaptation (RAA) framework, focusing on the business process as the primary unit of AI intervention. A conceptual system implementation context is proposed to illustrate RAA enablement at the process level and highlight the need for an RAI agentic architecture capable of measuring, analyzing, and forecasting AI's impact on business processes to support ethical, balanced innovation.

1 Introduction

The rapid adoption of artificial intelligence (AI) technologies is outpacing our ability to fully assess and respond to their long-term societal and economic impacts [1]. AI was initially expected to automate only repetitive, low-skill tasks, while the current landscape reveals a broader disruption, even highly skilled and complex human roles, such as software engineering, civil engineering, manufacturing, and finance, are being significantly affected [2]. Tasks that once required teams of multiple professionals can now be accomplished by a few, creating a disproportionately opportunistic environment that primarily benefits large corporations and those with monopolistic mindsets [2,3]. This disproportionately is fueled by well-equipped economic and technical resources available in large corporations. The rapid adoption of modern AI technologies, such as large language models (LLMs) and generative AI, has contributed to a paradigm shift in society. This shift has given rise to a phenomenon we define as socio-technological imbalance: *a widening gap between the swift implementation of AI technologies and the ability of society, its institutions, industry (Small and Medium Enterprises (SMEs) in particular), regulations, cultures, and economies to assess AI adoption impacts in the short and long run to adapt and govern these changes in a fair, ethical, transparent, traceable, sovereign, informed and inclusive way to intact and protect the fundamental rights of a human.* Although AI brings efficiency and innovation, it can simultaneously concentrate power and wealth, accelerate job displacement beyond economic recovery rates [4,5], and outpace the development of legal and ethical frameworks. As a result, this imbalance exacerbates inequality, restricts equitable access to AI benefits, and introduces ethical challenges that current institutions worldwide are not yet equipped to address [5]. For instance, institutions like OpenAI, originally founded as nonprofits to democratize AI, have shifted toward for-profit models. Furthermore, the right to be forgotten under Article 17 of GDPR [6] faces significant challenges in the context of LLMs, where user data can become deeply embedded in the model's parameters. Unlike traditional databases, LLMs often lack mechanisms to trace or remove specific personal data once it has influenced training [7]. To address this concern, techniques like machine unlearning have been proposed in [8] to early promise for selective data removal, though they are not yet fully practical for large models [9,10]. This shortcoming raises serious concerns about users' ability to revoke consent or erase their digital footprint from centralized AI platforms. These concerns lead to a disalignment of corporate interests with the public good and the principles of responsible innovation. In addition, unlike traditional rule-based systems, modern AI, especially LLMs with over 64 billion parameters, function as black boxes [11], making it challenging to predict their behavior and to ensure alignment with ethical standards. Furthermore, deploying modern AI systems such as LLMs or generative AI within an organization without thorough testing and impact assessments is analogous to releasing a vaccine without undergoing clinical trials; both actions carry significant risks due to the lack of evidence on safety, effectiveness, and unintended consequences.

The potential societal consequences arising from the rapid adoption of AI call for a cautious, evidence-based approach to the development and deployment of these technologies. There is a pressing need to empirically examine the assumptions surrounding the impact and ethical implications of AI. These assumptions must be rigorously tested and validated or refuted through data-driven research that integrates both social and technical perspectives. Hence, aligning with the aforementioned context, we must clearly define and understand:

- What constitutes "responsible" or "ethical" use of AI or "Responsible AI" (RAI)?

This question highlights the need to establish clear principles for a fair and responsible AI (RAI) ecosystem. Embracing core concepts such as traceability, transparency, and collaborative, win-win business models, particularly those that enable cross-domain value-chain integration [12], can help foster a more balanced and equitable AI landscape. Such responsible, AI-driven value chains have the potential to align the interests of diverse stakeholders and ensure a more inclusive and fair distribution of AI's benefits [13]. Hence, more research is needed to investigate the empirical evidence from real-world social contexts and technical prototypes, where modern AI applications are being or going to be adopted, which will be critical in shaping a transparent, traceable, and RAI adoption framework. One promising direction is the adoption of decentralized AI methods, particularly through the federation of AI agents. This federated model enables AI to be trained globally while operating closer to sources where data is generated [14], which is especially beneficial for SMEs related to distributed architectures, as it eliminates the need for sensitive data to be centralized. Such an approach not only enhances data privacy but also enables inclusive innovation across diverse economic actors [14]. Considering the importance of AI, addressing the ethical, social-technical, and privacy concerns, we investigate an important research question (RQ), stated as:

How can we design and develop a domain-specific RAI adoption ecosystem for a specific process within an organization that aligns with the principles of RAI?

Following this question, this study investigates how such an adoption ecosystem can be designed and governed at the process level to ensure ethical and sovereign use of AI, discourage opportunistic behavior, and support responsible implementation within specific operational domains. In doing so, the study aims to contribute to the evolution of AI as a responsible discipline, not solely focused on transformation, automation, or profit generation, but also capable of self-regulation. This includes monitoring its own behavior, collecting relevant data to inform policy-level decision-making, and promoting equitable growth. Ultimately, the study advances the philosophy of RAI that supports long-term socio-technological balance.

2 Methodology Framework

This concept paper adopts an exploratory research approach to conceptualize and model the RAI Adoption/Adaptation (RAA) framework at the local enterprise and organizational level [15]. The methodology is grounded in an extensive literature review of RAI principles, policy frameworks (e.g., EU AI Act, Organization for Economic Co-operation and Development (OECD) guidelines), and existing architectural models, combined with the authors' experiential insights from process automation in real-world contexts. Based on this foundation, a system architecture and mathematical model are designed to represent key RAI components, such as decision support for RAI principles and process-level RAA incorporation overview. The system model is used to visualize the impact of RAA in comparison to non-AI adoption scenarios, and reflects relevant impact factors in multiple dimensions.

To demonstrate the practical applicability of the proposed RAA framework, an implementation context is presented outlining the relevant toolchain and functional flow for AI integration at both the task and process levels. This implementation guides AI agents and an agentic design pattern to embody core RAI principles, such as capturing AI-generated input and output logs, ensuring the traceability of data and AI models' processing, and analyzing ethically sensitive events with human-in-the-loop mechanisms to alert relevant stakeholders for informed policy and/or decision-making. Together, these components operationalize transparency, accountability, and ethical oversight within enterprise workflows.

3 Relevant Literature Reflections

The antecedents of RAI governance encompass societal expectations and norms, organizational values, and the evolving principles of RAI within organizations [16]. Authors in [16] have defined RAI governance notion based on structural, relational, and procedural practices as a conceptualized framework that includes *a set of practices for developing, deploying, and monitoring AI applications in a safe, trustworthy, and ethical manner that ensures appropriate functionality of AI over the entire lifecycle*. According to [16] and ongoing efforts within the European framework for RAI, the foundational principles can be categorized into several key areas: accountability, diversity, non-discrimination and fairness, human agency and oversight, privacy and data governance, technical robustness and safety, transparency, and social and environmental well-being [17,18]. While these principles may appear under different names across various reports, they consistently reflect the same underlying ethical concerns. For instance, "transparency" is often discussed alongside "explainability," and "human agency and oversight" may be represented as "human review of technology." In contrast to much of the existing literature, we distinguish that transparency can be perceived from both a social and digital standpoint rooted in *accountability, access, and power*, while viewing explainability as a technical characteristic of AI models.

Transparency involves enabling stakeholders to understand *who made decisions, how, why, and with what consequences*, not just in terms of model mechanics, but in relation to the broader decision-making processes and institutional context. Explainability, by contrast, focuses on interpreting how the model processes inputs (e.g., datasets or knowledge fed into the neural networks) and generates outputs. Authors in [18] have explored AI governance through the lens of complex adaptive systems (CASs), conceptualizing AI as a socio-technological and adaptive system where people, policies, systems, data, AI, processes, and other elements co-evolve. This perspective emphasizes the need for holistic, organization-wide approaches that consider public values and societal concerns, advocating for joint accountability among all stakeholders.

The literature contains many similar terms to define RAI, as the concept is still in its research inception stage [16]. Therefore, the terms Trustworthy AI, Ethical AI, and Principled AI all converge on the same concept, contributing to the definition of *Responsible AI (RAI)*. This term is now widely accepted, and we also stem its use as RAI in this article and try to consolidate its definition from the EU AI Act perspective.

3.1 Towards RAI

The EU AI Act defines and operationalizes principles for trustworthy, human-centric, and RAI within the European legal framework. It emphasizes risk-based regulation, transparency, accountability, data governance, and ongoing monitoring for high-risk AI systems [19]. The Act also includes specific provisions for explainability, human oversight, and robust documentation throughout the AI lifecycle. These elements largely overlap with the concepts of RAI found in the literature; however, the EU AI Act takes it a step further by establishing binding legal obligations and enforcement mechanisms.

To integrate the EU AI Act's approach into RAI, we have added regulatory requirements and mapped the core concepts of RAI and descriptions in Table 1 to indicate where legal obligations apply. For instance, the concepts of "Accountability" and "Explainability" are not only best practices but also legal requirements for high-risk AI systems under the Act.

Formal Definition - RAI

RAI refers to the development, deployment, and use of artificial intelligence systems in a manner that ensures compliance with ethical principles, legal obligations, and societal values. It requires that AI systems are designed and operated to be trustworthy, transparent, explainable, fair, robust, secure, and respectful of fundamental rights, including privacy and non-discrimination. RAI also mandates ongoing human oversight, accountability for outcomes, and consideration of broader societal impact throughout the AI lifecycle.

This definition aligns with the EU AI Act's emphasis on a risk-based approach, where high-risk AI systems must meet strict governance, transparency, and accountability requirements, and where certain uses (such as social scoring or subliminal manipulation) are prohibited outright [20–23].

Table 1. RAI Definition and EU AI Act Requirements

Core Concept	Description/Sub-components	EU AI Act Requirements
Ethics	Fairness, accountability, non-discrimination, societal alignment	Mandates human-centric, ethical AI; prohibits unacceptable risk systems (e.g., social scoring, subliminal manipulation) [21,22]
Security/Safety	Robustness, reliability, soundness	Requires high-risk systems to be robust, secure, and undergo conformity assessment [20,21]
Privacy	Data protection, confidentiality	High-risk systems must use high-quality, unbiased data; strict data protection rules [20,21]
Explainability	Transparency, understandability	High-risk systems must be transparent and explainable [21,23]
Trustworthiness	User trust, confidence in AI outcomes	Promotes trustworthy AI through governance, oversight, and compliance [20,22]
Human-centered	Focus on user needs and societal impact	AI must be human-centric; requires human oversight for high-risk systems [21,22]

It is important to mention that *Transparency and Explainability* requirements are integral parts of the EU AI Act's RAI framework. Under *Transparency*, the EU AI Act imposes transparency obligations on AI system providers and deployers wherein users must be informed when interacting with AI, and the system's capabilities and limitations must be clearly communicated. Relevant information about the AI model design, data, and decision-making must be disclosed. These requirements apply to all risk categories, with stricter rules for high-risk and general-purpose AI systems. Similarly, for *Explainability*, the Act mandates that high-risk AI systems must be explainable, so that affected individuals should understand the logic behind AI decisions and have the right to clear explanations of how the AI functions and what influences its outcomes. These provisions aim to empower users, ensure accountability, and facilitate oversight, closely linking explainability to transparency.

3.2 Relevant Impacts of Rapid AI Adoption

Hartley et al. [24] examine the economic impact of Generative AI on the labor market, highlighting its role in enhancing productivity by enabling individuals to complete tasks more efficiently. Their findings suggest that younger, more educated, and higher-income individuals are more likely to adopt tools like LLMs

(LLMs), particularly in sectors such as customer service, marketing, and information technology. The study also notes that while LLMs can complement human labor by boosting productivity, they may also substitute certain job functions. These insights underscore the need for policymakers, businesses, and educational institutions to proactively address the challenges and opportunities presented by such technological advancements. Authors in [25] investigate the key barriers preventing workers from using modern AI tools like ChatGPT, despite their potential to significantly reduce task completion time. The major barriers identified are the need for training and employer-imposed restrictions. Other concerns include fears of job redundancy or over-dependence on technology; for instance, customer service workers fear replacement. At the same time, professionals in writing-intensive roles, such as journalism and teaching, responded by often avoiding ChatGPT as it diminishes their job satisfaction. The relevant impacts of modern AI adoption and related concerns are now being raised globally, as reflected in the World Economic Forum's Future of Jobs Report 2025, which forecasts that by 2030, approximately 92 million jobs will be displaced by AI and automation worldwide [26]. However, a critical question remains: *how can we define or transform AI into RAI from the very inception or after implementation?*

It is crucial to embed RAI principles at the inception of AI adoption, rather than attempting to retrofit them later, once the technology is already in use and its unintended impacts may have led to socio-technical imbalances. However, some damage control can still be avoided in later stages by making certain RAI adjustments in the current policies or governance processes. [27] consolidates diverse perspectives to establish a Trustworthy or RAI ecosystem, building on the European Union's framework and earlier studies. The study defines Trustworthy AI as adhering to seven technical requirements including *human agency and oversight, robustness and safety, privacy and data governance, transparency, diversity, non-discrimination and fairness, societal and environmental well-being, and accountability*—supported by three foundational pillars: *lawfulness, ethics, and robustness* (encompassing both technical and social dimensions). These criteria must be upheld throughout the AI system's entire lifecycle, ensuring compliance through rigorous auditing processes. The development of such RAI systems is not only a technical necessity but also a societal imperative to keep the socio-technological balance in place, critical for addressing current and future challenges in an ethically aligned manner.

4 System Model

RAI has been well-defined; however, there is a lack of literature on AI adoption at the process level. Processes are fundamental to organizations, where actors, operations, physical and digital systems, services, and technologies such as AI, Edge, and IoT converge, essentially forming socio-technological complex adaptive systems (CASs) [18]. Hence, this study contributes to this area by focusing on CAS-based processes for developing an *RAI Adoptable/Adaptable* ecosystem, which functions as an RAA System. Therefore, we formally define RAA at the process level, aligning to the definition of RAI governance at the organization level in [16] and the EU AI Act as follows:

"***RAA* or *Adopting(at inception time)/Adapting (after implementation) AI Responsibly* or *Responsible use of AI*** – *is broadly defined as an integrative, human-centered, and process-oriented approach that combines ethical decision-making, transparency, traceability, security, privacy, explainability, and accountability to ensure trustworthy and beneficial AI adoption with a win-win business model, and is capable of supporting/maintaining relevant socio-technological balance for all stakeholders.*"

RAA is increasingly characterized by a set of interrelated technical and organizational dimensions that shape both the design and governance of AI systems. A **process-oriented** perspective underpins this framework, emphasizing the structured integration of AI into the physical, digital, and social workflows of an enterprise. RAA operates not merely at the system level but across operational pipelines, embedding ethical oversight into each layer. The **ethical** dimension- encompassing fairness, accountability, and compliance with societal laws and norms, is treated not as an abstract guideline but as a measurable and auditable set of constraints guiding AI adoption/adaptation behavior, data handling, outcome, and relevant impact assessment.

An **integrative architecture** is essential, where socio-technical components, including AI governance, AI adoption design, impact assessment, and stakeholder engagement, converge into a coherent AI lifecycle management process. From a risk and compliance standpoint, **security and safety**, while human-in-loop, needs to be addressed through robustness testing, transparency, privacy preserving, traceability, and win-win situational design strategies that promote system trustworthiness and reliability in both social and technological environments. In this context, the **human-centered** AI adoption/adaptation design brings in social agents/actors (designers, users, regulators) as integral participants in feedback loops, drawing from CAS theory to emphasize co-evolution and system adaptability.

Privacy is a cornerstone of both RAA and regulatory frameworks like the EU AI Act. From an ethical standpoint, RAI emphasizes the protection of personal data through principles such as data minimization, informed consent, and privacy-by-design, ensuring that individuals retain control over how their data is used [28]. Complementing this, the EU AI Act embeds privacy safeguards into law, particularly for high-risk AI systems, by enforcing strict data governance, transparency, and risk mitigation measures [21]. Together, these approaches aim to uphold individual rights, foster trust, and promote safe, privacy-respecting innovation with AI.

Similarly, **transparency and traceability** are achieved by implementing process control, audit logs, AI adoption decision documentation (e.g., model cards, datasheets for datasets), and regulatory reporting mechanisms. The relevant socio-technical enablers contribute to **explainability** aspect during the AI adoption/adaptation process, wherein AI models and their impacts can be made interpretable through causal inference methods, or inherently interpretable enterprise architectures subject to AI automation, especially in high-stakes domains. In this context, SHAP and LIME are two popular model-agnostic explanation methods that help interpret the predictions made by complex (often black-box) AI models like random forests, XGBoost, and neural networks.

A **win-win business model** refers to a strategy or arrangement where all parties involved benefit. Therefore, in the current context, the push for a **win-win business model** introduces a socio-economic layer, wherein AI value creation aligns with stakeholder interests through multi-objective optimization and impact assessment frameworks. Furthermore, the **socio-technological balance** can be maintained by designing governance structures, during the pre and post adoption of AI in the given process, that address the interdependencies among socio-technological actors and entities that include humans, systems, services, data, algorithms, and institutional norms, often supported by tools like AI impact assessments, algorithmic audits, and continuous monitoring systems. Ultimately, **trustworthiness** emerges as a systemic property, not just of algorithms, but of the entire AI ecosystem built through transparent processes, robust validation, ethical alignment, and continuous engagement with users, stakeholders, and regulators.

It is important to mention that there is a lot of ongoing research that aims to clarify terminology and develop actionable frameworks for RAI implementation. Therefore, there are a lot of interrelated constructs. For example, Ethical AI, Trustworthy AI, and Human-Centered AI are closely related but distinct concepts, each contributing to the broader idea of RAI. Trustworthiness is often seen as an outcome of responsible practices, while explainability and privacy are foundational requirements. In addition, there is an overlap and ambiguity in terms like fairness, transparency, accountability, and sustainability, which are frequently included but are often subsumed under broader categories such as ethics or security.

4.1 RAA Impact Assessment

As organizations increasingly integrate AI into critical decision-making systems, it becomes necessary to assess not just the performance of these models but also the broader socio-technical impacts of their deployment. Therefore, RAA, aligned with its definition based on RAI, emphasizes the importance of aligning AI systems with ethical principles, legal requirements, and societal values. To evaluate the extent and effectiveness of RAA, we require well-defined RAA models that can quantify its impact across multiple dimensions such as fairness, accountability, transparency, privacy, and robustness. These models must move beyond traditional accuracy-centric metrics to incorporate a multi-perspective view that reflects both the benefits and potential harms of AI deployment. By formalizing these definitions, we can systematically measure and compare the outcomes of AI systems deployed with and without RAI principles, enabling data-driven insights, continuous improvement, and trustworthy governance. We assume $F(x)$ represents the RAI index and can be mathematically modelled as,

$$F(x) = [f_1(x), f_2(x), \cdots, f_n(x)] \quad \forall \; x \in X, \tag{1}$$

where $f_n(x)$ and X represent the different RAI dimensions and feasible solution space, respectively. To optimize the vector-valued function in Eq. 1, where each function represents a distinct RAI-related goal. For instance, $f_1(x)$ might measure model accuracy, $f_2(x)$ a fairness metric, $f_3(x)$ the privacy loss (e.g., differential privacy parameter ϵ), and $f_4(x)$ an interpretability and explainability. These objectives are subject to constraints such as regulatory, ethical, or technical limitations within a feasible solution space X. To identify a Pareto optimal solution, where improving one objective cannot occur without compromising another, can be applied.

To maximize the RAI, first, we need to find the optimal solutions for each organizational dimension. Starting with accuracy $f_1(x)$, which can be calculated with a ratio of true positives and true negatives to all four possibilities (true positives (TF), true negatives (TN), false positives (FP), and false negatives (FN)). However, for G number of demographic or social subgroups, accuracy can be represented mathematically as,

$$RAI_{Accuracy} = \frac{1}{G} \sum_{g=1}^{G} \frac{TP_g + TN_g}{TP_g + FP_g + TN_g + FN_g} \tag{2}$$

The fairness can be expressed as the expectation of the difference between two states as a structural causal model. In order to represent mathematically, we consider one decision variable with two possible states (0 and 1), as follows:

$$f_2(x) = \mathbb{E}[Y_{O=0} - Y_{O=1}] \tag{3}$$

The fairness metrics can be based on demographic and predictive parity, individual and resource fairness, and equal opportunity, and can also be expressed as a loss function. Similarly, privacy can be modelled by using a privacy loss function, generally based on Differential Privacy (DP), which provides a formal guarantee that individual data points have limited influence on the output of an algorithm (A). Assume an algorithm A is said to be (ε, δ)-differentially private if for any two datasets D and D' differing by only one record, and for any output set S,

$$Pr[M(D) \in S] \leq e^{\varepsilon} \cdot Pr[M(D') \in S] + \delta, \tag{4}$$

where ε is the privacy budget, and the smaller the value, the stronger the privacy. The probability of breaking the privacy guarantee is denoted by δ, ideally close to zero. Privacy can be normalized as $f_3(x) = 1 - \varepsilon(x)/\varepsilon_{max}$, and upper limited with a threshold value to enforce privacy guarantees.

The explainability is commonly assessed through SHAP/ LIME coverage, feature attribution, and explanation simplicity. For better explainability, the higher SHAP/LIME coverage, shorter, and more focused explanations (yielding lower entropy) are desired.

Finally, after finding the optimal regions for each organizational dimension, the objective function of RAI $(F(x))$ can be written as,

$$F(x)_{max} = \sum_{i=1}^{n} \lambda_i f_i(x) \quad \forall \quad i \in \{1,2,3,4\} \tag{5}$$

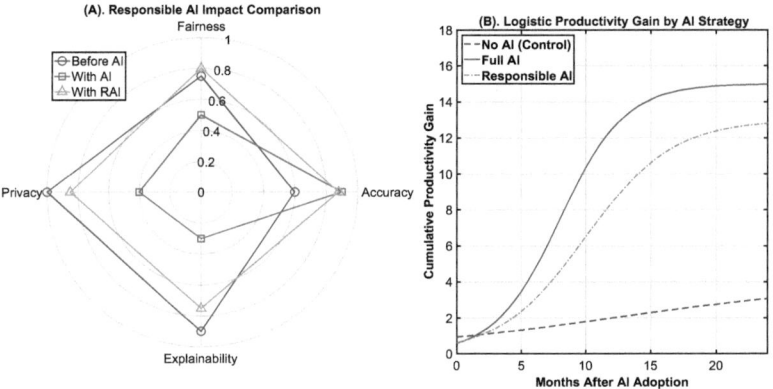

Fig. 1. Comparisons of RAI

By formally modeling these dimensions as a vector-valued function, organizations can systematically identify trade-offs, optimize for Pareto-efficient solutions [29], and ensure their AI systems align with ethical standards, legal requirements, and societal expectations. This approach not only promotes trustworthy and transparent AI governance but also empowers continuous improvement and accountability, fostering greater public confidence in AI-driven decision-making.

4.2 Impact on Organizational Dimensions with RAA

To holistically evaluate the benefits of adopting RAI, it is important to consider both its ethical impact and long-term productivity gains. While earlier sections compared individual metrics and performance trends, this final impact function integrates those insights into a unified perspective. By balancing accuracy, fairness, privacy, and explainability alongside projected productivity, we can better understand the comprehensive value of implementing AI with responsible safeguards. Based on fictitious data, the two graphs presented in Fig. 1 provide a comprehensive analysis of the impact of AI adoption under different governance strategies, such as traditional (human-only) processes, AI without oversight, and RAI with ethical safeguards. Together, they explore both the qualitative trade-offs in AI decision-making and the quantitative productivity gains over time.

The first graph, Fig. 1.A evaluates four critical metrics, Accuracy, Fairness, Privacy, and Explainability, across the three scenarios. The chart reveals that while AI without safeguards achieves the highest accuracy, it performs poorly in fairness, privacy, and explainability. In contrast, the traditional (pre-AI) process excels in privacy and explainability while lacking accuracy. The RAI approach strikes a balanced profile: it maintains high accuracy while significantly improving fairness, privacy, and explainability compared to unsupervised AI. This demonstrates that responsible governance mechanisms can mitigate AI's ethical shortcomings without greatly sacrificing performance. The second part, Fig. 1.B models the cumulative productivity gain over a 24-month period following AI adoption. The logistic growth curves highlight how each approach performs over time. As expected, AI without constraints yields the fastest and largest gains. However, the RAI scenario closely follows with only a modest delay in reaching peak performance. In contrast, the no-AI (human-only) process shows limited improvement, reflecting a missed opportunity for innovation. The RAI curve demonstrates that it's possible to achieve substantial productivity benefits while still enforcing ethical safeguards.

In combination, these two visualizations tell a compelling story: RAI delivers nearly all the performance benefits of unconstrained AI, while preserving the fairness, transparency, and trustworthiness that are essential for long-term success. Organizations that adopt AI responsibly are positioned not only to innovate but to do so in a way that aligns with ethical standards, regulatory expectations, and public trust.

5 RAA Framework

RAI Adoption in RAA refers to the initial phase of integrating AI automation into existing processes, while RAI Adaptation describes the subsequent operational phase, where AI has been embedded and is actively functioning within those processes. The overall framework for adopting and adapting RAI at the process level is illustrated in Fig. 2. Figure 2a presents RAA Framework, which consists of 8 components or functional blocks, named and explained as follows:

1. **Organizational or Enterprise Settings:** Refers to the broader domain-level boundaries, including assets (such as data, systems, and services), resources (digital, physical, or social), processes, operations, values, and value chains that define the functioning of the organization.
2. **Physical Infrastructure:** This represents the physical, mechanical, and digital infrastructure necessary to support domain-specific operations, particularly for implementing AI services.
3. **Digital Systems:** This encompasses digital services, digital twins, and shadow systems within the domain that support the operations of physical infrastructure. These systems manage data lifecycle tasks and related processing activities, serving as primary candidates for AI-driven automation [30].
4. **Applications and Interfaces:** This refers to the software applications built on top of digital system interfaces, including AI-based centralized, localized,

or federated services. These are instrumental in enforcing RAI principles and ensuring compliance with legal and ethical standards.
5. **Social Actors and Implications:** This represents stakeholders within the organizational domain, such as employees, management, customers, end-users, supply chain partners, and vendors, who are directly or indirectly impacted by the adoption or integration of AI technologies.
6. **AI Ecosystem:** This consists of AI models, training processes, toolchains, and orchestration mechanisms required to deploy AI within domain-specific processes. Here, processes are understood as collections of distributed tasks subject to AI-based automation.
7. **RAI Philosophy:** This provides the principles, guidelines, governance mechanisms, and best practices of RAI, as informed by frameworks such as the EU AI Act and broader ethical and societal considerations. It aims to maintain a socio-technological balance during AI adoption or integration.
8. **RAI Enablement Methods:** This refers to a combination of business and technological strategies, including qualitative and quantitative methodologies, to monitor AI life cycle events and related systems inputs, processes, and outputs. These methods help assess impact, inform policy development, and support data-driven decision-making aligned with RAI principles, fostering sustainable growth within the organizational context.

This framework is designed to support the integration of AI in a manner that aligns with RAI principles across various organizational processes. AI adoption typically occurs at the task level within a given process, with each process comprising multiple tasks within a specific domain. The primary objective is to ensure RAI usage throughout both the pre-implementation (adoption) and post-implementation (adaptation) phases. The aim is to achieve the RAI-compliant state eventually for a given process along with AI-optimized gain while ensuring ethical, transparent, and socially responsible RAI adoption, as shown in Fig. 2b. Such configurations typically result from implementing RAI enablement methods that consist of combinations of qualitative or quantitative methods through stakeholder consultation and analysis targeting task-level suitability during AI integration. From the perspective of RAI maturity and integration, as shown in the Fig. 2b, four types of process transition levels has been identified:

1. **Non-AI (NAI) Processes:** These are existing processes that currently do not utilize any AI components, even though AI integration might be possible. This may be due to lack of awareness, expertise, or impact assessment. Such processes continue to operate using traditional methods, typically involving human intervention, multiple systems, and manual services. These processes are usually subject to applying AI automation in the near future.
2. **Partial-AI (PAI) Processes:** These processes have selectively adopted AI to certain tasks while others remain non-automated. This results in a hybrid process flow combining both AI and non-AI operations.
3. **Full-AI (FAI) Processes:** In this case, AI is fully integrated across all tasks within the process, with no manual or non-AI components remaining except the input and output controls.

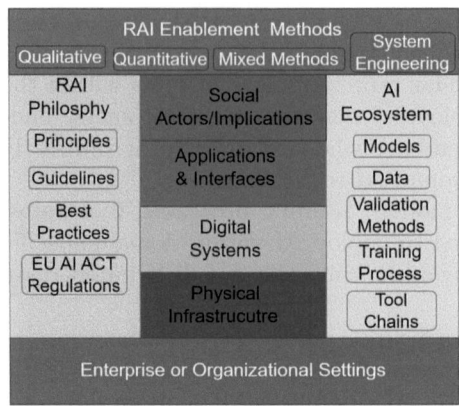 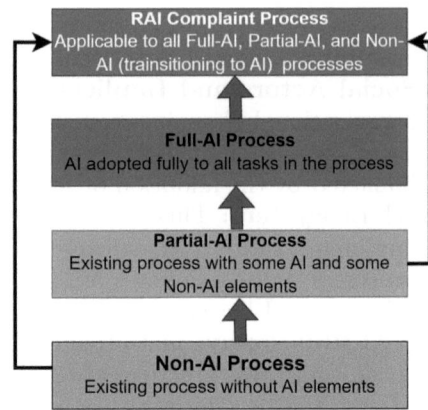

(a) RAI Adoption/Adaptation (RAA) Framework

(b) Process Transition Model (Non-AI to RAI): Types and States

Fig. 2. High-level RAI Adoption/Adaptation Framework

4. **RAI Compliant (RAIC) Processes:** These processes apply to all processes, i.e., NAI (subject to AI adoption in near future), FAI or PAI processes. The decision to transition the path towards RAI from a certain level can be based on comprehensive business modeling, cost-benefit analysis, resource availability, and a commitment to comply with RAI principles, as exemplified in the process transition model presented in Sect. 2b.

The overarching objective of this architecture is to guide all processes toward becoming *balanced RAIC processes* as shown in Fig. 2b, in line with RAI principles. This helps organizations maintain an equilibrium between technological advancement and socio-impact considerations at the process level.

5.1 System Implementation Context

Figure 3 illustrates the organizational domain context, which comprises multiple processes, each consisting of several tasks. From a generalization perspective, these processes vary in type NAI, PAI, FAI, and RAIC as described earlier in relation to AI adoption, Fig. 2b. Typically, AI is integrated at the PAI and FAI levels without adhering to RAI principles or regulatory compliance. To align with the EU AI Act, existing processes must be adapted to incorporate RAI principles. Ideally, RAI should be embedded during the transition from NAI to PAI or FAI. However, if processes are already implemented, they must be retrofitted to conform to the RAI framework. This raises a key question: *How can RAI be operationalized at the process level?* We propose introducing task-level AI agents referred to as Light Weight Agents (LWA), designed to monitor and measure specific metrics at the task level to quantify the impact of AI automation, eventually at the process or domain level. These agents collect relevant data at

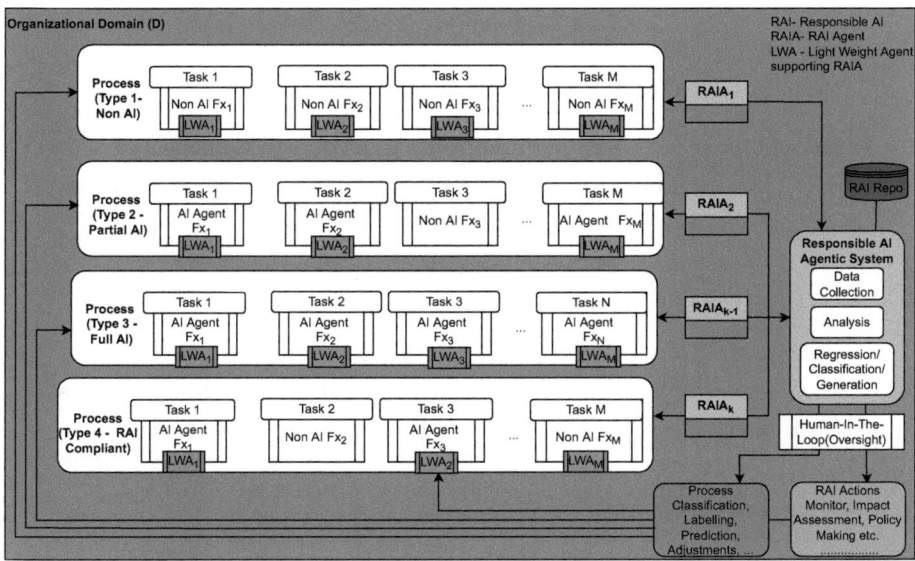

Fig. 3. Process-oriented RAI Enablement Architecture.

the input, processing, and output stages using RAI enablement methods. The impact can be quantified using metrics as exemplary outlined in the Sect. 4.1. Each task-level agent operates locally and contributes to a federated framework, enabling the assessment of RAI impact at the process and, subsequently, the domain level. To facilitate this, the architecture introduces an RAI agent at the process level and utilizes LWA at the task level, which operate in a federated manner across intra-process (using LWA) and inter-process (using RAIA) levels, while maintaining proximity to the supervised processes. Their primary function is to continuously collect and analyze data—both quantitative (e.g., system logs, performance metrics) and qualitative (e.g., human feedback, text and video analysis)—to ensure the responsible and accountable deployment of AI.

To support this, clearly defined RAI metrics are essential for consistent monitoring and evaluation. Qualitative inputs can also be translated into quantitative indicators (e.g., using Likert Scales, Rating Scales, qualitative analysis with coding schemes, etc.) to aid in assessing RAI compliance. To support local validation of the RAI adoption framework, we propose a modular RAA implementation stack, outlined in Table 2. This stack emulates real-world AI integration across various process types and is structured to encompass simulation, AI task execution, ethical auditing, monitoring, and visualization. While the table is largely self-explanatory, following additional insights are provided on the validation mechanisms that underpin the effectiveness and adaptability of the proposed framework.

Table 2. RAA Implementation Stack

Functional Layer	Purpose/Tools/Features/Components
Process Orchestration	To model process and task level workflows associated to NAI, PAI, FAI and RAI prcoess types. Tools: N8N, Camunda, Apache Airflow, BPMN engine; Features: - Process-type tagging (NAI, PAI, FAI, RAIC) - Task-level granularity for RAI compliance - Integration hooks for AI and RAA modules
AI Task Agent	To plug AI modules into tasks (e.g., prediction, classification, decision-making); Components: - Inference Agent: Integrates ML models (e.g., TensorFlow, PyTorch, Hugging Face models) - Feedback Agent: Captures user/system responses - Audit Logger: Records input-output mappings and model confidence
RAI Enablement	To ensure ethical, legal, and compliance-based monitoring; Components: - RAI Metrics Engine: Computes fairness, transparency, explainability, etc. scores - Bias & Drift Detectors: (e.g., IBM AI Fairness 360, What-If Tool, Amazon SageMaker Clarify) - Explainability Tools: SHAP, LIME, Captum, etc. - Ethics Scoring Module: Combines qualitative-quantitative feedback (e.g., via Likert, fuzzy mapping, or qualitative coding scheme, etc.)
Data Collection & Monitoring	To collect operational and impact-level data at runtime; Tools: - Log-based Observability (OpenTelemetry, Prometheus) - Qualitative Inputs (e.g., survey injectors, video feedback) - Event Stream Analyzers (Apache Kafka, Flink) - KPI Dashboards (Grafana, Kibana)

(*continued*)

Table 2. (*continued*)

Functional Layer	Purpose/Tools/Features/Components
Federated RAI Agentic Network	To support scalable, decentralized RAI impact monitoring, learning and model training; Features: - Each task/process hosts a lightweight RAI Agent - Agents report to process-level RAI aggregators or Agents to produce domain-level analysis, classification, or prediction of RAI impact and compliance. - Federated architecture for intra and inter-process level RAI features/metrics learning with human in the loop
Human-in-the-Loop (HITL) Interface	To review the decision or prediction outcomes of the AI enables the service outcome. Tools: Qualitative, Quantitative, and Mixed methods for input and output analysis w.r.t RAI philosophy
Process Simulation And Testbed	To emulate transitions from NAI → PAI → FAI → RAIC using synthetic or real datasets based on emails, system logs, and customer or stakeholder feedback, etc.; Features: - Simulate synthetic or real process workflows - A/B testing for RAI adaptation/adoption - Scenario-based RAI compliance monitoring and related functional testing

Validation Metrics and Monitoring. The following exemplary metrics can be constituted at the process level and can be collected during RAA process for AI impact assessment for comparing NAI, PAI, FAI, and RAI scenarios:

- Proportion of tasks handled by AI vs non-AI methods.
- Execution time and efficiency gains.
- Fairness and bias indicators or kind of scoring entities.
- Return on Investment (ROI) estimates.

RAA Modelling/Simulation Over Time. Based on real data or a simulated environment, process modelling with arbitrary values, iterative simulations, emulation, or prototype can be developed to gradually transition processes from Non-AI to RAIC process types, guided by impact assessments. RAA Agents can be deployed alongside the Non-AI processes to continuously collect data, thereby helping to establish baseline metrics at the process level. These baselines are essential for evaluating the effectiveness of, and informing decisions around, AI adoption and adaptation strategies. They also support ongoing alignment with RAI principles by ensuring that all relevant stakeholders are included in the decision-making process.

Implementation Insights. We can create a lightweight AI agent, in combination with other technologies, following the AI Agentic design pattern, to monitor the **input, output, and processing time** of a given **AI model performing a task**. This type of agent can act as a wrapper or observer around (in a non-invasive way, e.g., SHAP or LIME methods) the AI model and is often used for logging, auditing, or performance analysis.

5.2 Use Case Relevance: RAI For Industrial Customer Support Automation

This section maps a real-world industrial use case scenario of *industrial customer support automation* where the AI ecosystem, guided by RAI principles, can be applied to achieve goals such as transparency, traceability, and accountability, particularly when integrated into enterprise processes or pipelines. While the specific implementation may vary across domains due to diverse process requirements, several components of the proposed framework remain broadly applicable and domain-agnostic.

- **Fairness and Bias Mitigation:** In implementing the AI system for automating customer support ticket classification, fairness across user demographics and categories needs to be assessed. Model training data needs to be audited to ensure balanced representation across different issue types, urgency levels, and customer profiles. Post-deployment monitoring includes fairness metrics such as disparate impact and equal opportunity to prevent discrimination.
- **Transparency and Explainability:** The AI system can incorporate explainable models like SHAP (SHapley Additive exPlanations) to provide human-readable justifications for each classification. This helps human agents validate AI decisions and builds trust in the system.
- **Accountability and Oversight:** Human Oversight or Human-in-the-loop (HITL) is needed to ensure that human agents review edge cases or low-confidence predictions by PAI, or FAI processes. An AI operations team leveraging AI tool chain can be designated for continuous auditing, retraining, and performance evaluation of the system.
- **Social Impact:** One approach to measuring the social impact of AI adoption is to assess productivity gains, workforce displacement effects, and opportunistic behavior shifts. For instance, automation may lead to the reduction of a few or even a significant number of Full-Time Equivalents (FTEs). It is therefore critical to reassess impacted roles during both the adoption and adaptation phases using appropriate simulation or modeling techniques. This allows organizations to proactively plan reskilling or upskilling strategies to

ensure a fair and just transition for all stakeholders. Additionally, organizations should either adhere to or establish ethical workforce transition policies aligned with RAI principles, ensuring that technological advancement does not come at the cost of workforce well-being.
- **Privacy and Data Governance:** In accordance with the EU AI Act, AI models must ensure transparency in data usage and uphold ethical standards, particularly in relation to fundamental rights. To support this, RAA at the process level should conduct rigorous and continuous monitoring to verify that any user data used for training is anonymized and fully compliant with GDPR and internal data governance policies. Furthermore, access to training data and model predictions must be traceable, transparent, and subject to regular audits by relevant stakeholders. This ensures accountability, prevents misuse of data sovereignty, and promotes consistent regulatory conformity.
- **Continuous Monitoring and Feedback Loops:** The AI system should include performance drift detectors and RAI metrics dashboards. A feedback loop from customer satisfaction surveys and human agent overrides should be used to improve model quality over time.

6 Conclusion

AI technologies are being increasingly adopted across industries, with a focus on integration into both repetitive and highly skilled domains. This trend is reshaping the socio-economic and socio-technological landscape. AI-driven automation promises greater efficiency and cost savings, but it also introduces complex ethical, governance, and equity challenges, particularly when the benefits disproportionately favor large corporations, sidelining smaller players such as SMEs. As a result, a phenomenon of socio-technological imbalance is expected to emerge over time. To address this imbalance, the concept of RAI has emerged as a critical paradigm. RAI emphasizes principles such as fairness, transparency, traceability, sovereignty, and governance. The literature review reveals that while the concept of Responsible AI is well-defined, it is often referred to by other terms such as Trustworthy AI or Governance AI. The EU AI Act has formally documented principles, actions, and legal obligations related to RAI and Trustworthy AI. However, the practical implementation of RAI is still in its early stages of research. To advance the theoretical understanding of RAI, a mathematical model has been proposed to quantify RAI principles, allowing for comparison between RAI-compliant and non-compliant scenarios. Building on this, the present research proposes a domain-specific framework for RAI adoption and adaptation at the organizational process level. It introduces a process transition model based on the level of AI adoption within specific processes. In addition, the study presents a system implementation context leveraging decentralized architectures, such as federated AI agents and the AI agentic design pattern. Finally, this work gives the foundation for designing RAI and adapting AI to RAI ecosystems that are not only intelligent and efficient but also responsible, driven by ethical, inclusive, and well-governed practices to ensure that technological progress aligns with broader societal well-being and contributes to a sustainable socio-technological balance.

References

1. Li, B., et al.: Trustworthy AI: from principles to practices. ACM Comput. Surv. **55**(9), 1–46 (2023)
2. Septiandri, A., Constantinides, M., Quercia, D.: AI and the economic divide: how artificial intelligence could widen the divide in the US. EPJ Data Sci. **14**(1), 33 (2025)
3. Rockall, E.J., Tavares, M.M., Pizzinelli, C.: AI adoption and inequality. In: IMF Working Paper 2025/068, International Monetary Fund (2025). Accessed 14 Aug 2025
4. Challoumis, C.: The future of work – AI's impact on employment and the economy. In: Proceedings of the XVIII International Scientific Conference (2024). Accessed 29 May 2025
5. Bircan, T., Özbilgin, M.F.: Unmasking inequalities of the code: Disentangling the nexus of AI and inequality. Technol. Forecast. Soc. Change **211**, 123925 (2025)
6. General Data Protection Regulation (GDPR): – article 17: Right to erasure ('right to be forgotten'). https://eur-lex.europa.eu/eli/reg/2016/679/oj, 2016. Regulation (EU) 2016/679 of the European Parliament and of the Council
7. Blanco-Justicia, A., et al.: Digital forgetting in large language models: a survey of unlearning methods. Artif. Intell. Rev. **58**(3), 90 (2025). https://doi.org/10.1007/s10462-024-11078-6
8. Hine, E., Novelli, C., Taddeo, M., Floridi, L.: Supporting trustworthy AI through machine unlearning. Sci. Eng. Ethics **30**(5), 43 (2024)
9. Hyeonsu, L., Yang, H.J.: Machine unlearning: a survey on principles and challenges. In: Proceedings of the KICS Conference, pp. 1360–1361 (2023)
10. Liu, S., et al.: Rethinking machine unlearning for large language models. Nat. Mach. Intell. **7**, 1–14 (2025)
11. Ramlochan, S.: The black box problem: opaque inner workings of large language models. Prompt Engineering (2023). https://promptengineering.org/the-black-box-problem-opaque-inner-workings-of-large-language-models/. Accessed 18 July 2024
12. Singh, P., Beliatis, M. and Presser, M., et al.: Data-driven IoT ecosystem for cross business growth: an inspiration future internet model with dataspace at the edge. In: INTERNET 2024: The Sixteenth International Conference on Evolving Internet, ISBN: 978-1-68558-133-6 (2024)
13. Blair Attard-Frost and David Gray Widder: The ethics of AI value chains. Big Data Soc. **12**(2), 20539517251340604 (2025)
14. Yang, Q.: Toward responsible AI: an overview of federated learning for user-centered privacy-preserving computing. ACM Trans. Interact. Intell. Syst. (TiiS) **11**(3–4), 1–22 (2021)
15. Aithal, P.S., Aithal, S.: New research models under exploratory research method. In: Paul, P.K., et al. (eds.) A Book "Emergence and Research in Interdisciplinary Management and Information Technology", pp. 109–140. New Delhi Publishers, New Delhi (2023)

16. Papagiannidis, E., Mikalef, P., Conboy, K.: Responsible artificial intelligence governance: a review and research framework. J. Strateg. Inf. Syst. **34**(2), 101885 (2025)
17. Cannarsa, M.: Ethics guidelines for trustworthy AI. In: The Cambridge Handbook of Lawyering in the Digital Age, pp. 283–297 (2021)
18. Janssen, M.: Policy and Society, p. puae040 (2025)
19. European Commission: AI act - shaping Europe's digital future (2024). Accessed 30 June 2025
20. The act texts | EU artificial intelligence act (2024). Official AI Act documents and texts
21. Smuha, N.A.: Regulation 2024/1689 of the Eur. Parl. & Council of June 13, 2024 (Eu Artificial Intelligence Act). Technical report
22. European Commission: AI act | shaping Europe's digital future (2025)
23. Bird & Bird: European union artificial intelligence act: a guide (2024)
24. Hartley, J., Jolevski, F., Melo, V., Moore, B.: The labor market effects of generative artificial intelligence (2024). Available at SSRN
25. Humlum, A., Vestergaard, E.: The unequal adoption of ChatGPT exacerbates existing inequalities among workers. Proc. Natl. Acad. Sci. **122**(1), e2414972121 (2025)
26. World Economic Forum: The future of jobs report 2025, 2025. Accessed 6 June 2025
27. Díaz-Rodríguez, N., Del Ser, J., Coeckelbergh, M., De Prado, M.L., Herrera-Viedma, E., Herrera, F.: Connecting the dots in trustworthy artificial intelligence: from AI principles, ethics, and key requirements to responsible AI systems and regulation. Inf. Fus. **99**, 101896 (2023)
28. Singh, P., Meratnia, N., Beliatis, M.J., Presser, M.: Navigating the international data space to build edge-driven cross-domain dataspace ecosystem. In: Presser, M., Skarmeta, A., Krco, S., González Vidal, A. (eds.) International Summit on the Global Internet of Things and Edge Computing, pp. 151–168. Springer, Cham (2024). https://doi.org/10.1007/978-3-031-78572-6_10
29. Lin, X., et al.: A pareto-efficient algorithm for multiple objective optimization in e-commerce recommendation. In: Proceedings of the 13th ACM Conference on Recommender Systems, pp. 20–28 (2019)
30. Singh, P., Haq, A.U., Beliatis, M., et al.: Meta standard requirements for harmonizing dataspace integration at the edge. In: 2023 IEEE Conference on Standards for Communications and Networking, pp. 130–135. IEEE (2023)

Open Access This chapter is licensed under the terms of the Creative Commons Attribution 4.0 International License (http://creativecommons.org/licenses/by/4.0/), which permits use, sharing, adaptation, distribution and reproduction in any medium or format, as long as you give appropriate credit to the original author(s) and the source, provide a link to the Creative Commons license and indicate if changes were made.

The images or other third party material in this chapter are included in the chapter's Creative Commons license, unless indicated otherwise in a credit line to the material. If material is not included in the chapter's Creative Commons license and your intended use is not permitted by statutory regulation or exceeds the permitted use, you will need to obtain permission directly from the copyright holder.

Distributed and Trusted Access to Data Spaces' Products Employing Sovity and Data Fabric

Matilde Julian[1], Miguel Ángel Esbrí[2], Ignacio Lacalle[1](✉), Lucía Cabanillas[3], Rafael Vaño[1], and Carlos E. Palau[1]

[1] Communications Department, Universitat Politècnica de València (UPV), Valencia, Spain
{majuse,iglaub,ravagar2}@upv.es, cpalau@dcom.upv.es
[2] EVIDEN, Madrid, Spain
miguel.esbri@eviden.com
[3] Telefónica I+D, Madrid, Spain
lucia.cabanillasrodriguez@telefonica.com

Abstract. The current data landscape presents several challenges in data governance and integration. This paper presents the integration of the Data Fabric implemented in the Horizon Europe project aerOS, which integrates and unifies data available in the IoT-Edge-Cloud continuum, with the Data Space implemented in the RE4DY project in order to extend the capabilities of the Data Fabric by enabling the definition fine-grained data access policies and data monetization. This integration would promote collaboration with other participants in the Data Space based on the reuse of data assets from the Data Fabric by sharing them in a trusted environment and the integration of new data or services with the Data Fabric, thus contributing to the creation and integration of new data ecosystems.

Keywords: Data Fabric · Data Space · Trusted Connector · Federation

1 Introduction

The evolving challenge of distributed digital data management -as, in general, all computing landscape- has experienced a proliferation of heterogeneous sources, exposing data in different formats and structures from varied locations, but also through different data access protocols depending on the technology used to implement each data source.

The notion of a meta operating system (Meta-OS) that spans multiple technological domains that form the continuum -from IoT sensors to edge locations, and up to the cloud- increases the complexity, but offers a handy conceptual framework for distributed data management. Data sources are not only located in some physical locations, such as IoT sensors, but they are also spread over multiple physical and virtual locations across the different domains [16].

Besides, being the continuum a highly changing environment, as new data sources may become available, move to other domains, and even disappear. Such an intricate and heterogeneous data landscape introduces several challenges in two main aspects: i) data analysis and ii) data governance.

Vertical services, such as ML applications, that aim to consume and analyse these data to realize their business capabilities, would have to deal with such a complex and heterogeneous data landscape. They would also need to implement specific mechanisms for collecting, processing, and consuming the data of interest. Moreover, services shall correlate and combine data from multiple data sources. All these data analysis processes entail an extremely time-consuming effort that, in addition to having data engineering skills, requires a deep understanding of the available data. Similarly, data governance processes must be able to cope with this diversity of data in a dynamic environment. The security and privacy teams must ensure that data are properly classified and protected, accessed only by authorised consumers, and used for a specific reason. Keeping track of all these activities calls for a holistic view of the available data and how they are exchanged within the continuum.

In this work, authors combine the concept and technical depiction of two intricate solutions (Data Fabric and Data Spaces) to make a step forward in the establishment of a trusted data management framework on top of a distributed, multi-stakeholder computing continuum. Making use of the already available results of two research projects, the paper sketches the implementation of a novel system for achieving better data access control and traceability based on open European software.

The rest of the paper is structured as follows. Section 2 briefly outlines the context of the research projects under study, defining key concepts such as Data Fabric, Data Spaces and trusted connectors. Section 3 digs into the design of the new, proposed solution, dissecting the system into connected modules. It also exposes how it would be configured in an end-to-end integrated scenario. Finally, the paper concludes in Sect. 4 with some reflection and future working lines.

2 Related Background

2.1 Context

A Data Space is a distributed and trusted framework that enables secure exchange and trading of data assets among its participants based on specific rules and policies while ensuring data sovereignty.

The Horizon Europe project RE4DY [9] aims to enhance data management capabilities in the industrial sector through the creation of distributed data-driven value networks that support all phases of the product lifecycle. RE4DY emphasizes the importance of decentralization, interoperability and collaboration. This implies that data integration, discovery, governance, curation, and orchestration must be addressed [13]. Following its objectives, RE4DY is aligned with the Data Space design principles [15].

On another note, the Horizon Europe project aerOS [11] aims, among others, to support needs of data producers/consumers towards collection, processing, storage, and distribution of actionable data in the computing continuum. The so-called Data Fabric paradigm defines a new data infrastructure architecture that enables integrating data from heterogeneous data sources, structuring a semantic knowledge layer composed of a catalogue of products that stick to a specific ownership and access schema [14].

2.2 Distribution: aerOS Project

To crystallize the distribution of data in a true sense of continuum, it is necessary to depict a unified view of the data, regardless of their original location, format, or data source technology. To this end, aerOS relies on a homogenization layer -based on semantics- that abstracts data consumers from the underlying complexities of the continuum and exposes the data through a standard interface. Precisely, this is the idea behind the data fabric paradigm; however, aerOS goes beyond that, by extending the data fabric throughout the IoT-Edge-Cloud continuum. The aerOS Data Fabric aims to realise a one-stop-shop for data consumers to easily find, understand, and access any data available in the continuum; whereas data governance is provided with a complete view of how these data could be used and by whom.

2.3 aerOS Data Fabric

The Data Fabric in aerOS is a metadata-driven architecture that automates the integration of data from heterogeneous sources and exposes the data through a standard interface. The Data Fabric transforms the raw data of the providing domain into a data product that follows a standard data model, and the resulting data product can be shared with consuming domains through the standard interface of the Data Fabric. The aerOS Data Fabric includes mechanisms supporting data governance procedures for a traceable and responsible usage of data in the continuum. Keeping track of the provenance of the data, knowing how and where data are generated in the continuum, and who are consuming the data, are essential requirements towards maintaining full control over the data. In addition, it covers the conception of all data available in a continuum as a single box that can be queried and will forward the proper information. Technologically, aerOS has developed a Data Fabric implementation based on Orion-LD Context Broker, OpenLDAP, Morph-KGC, the library rdflib and the security supporting tools KeyCloak and KrakenD [12].

To round up the circle, Data Fabric represents the concept of a unified architecture, where computing capabilities are distributed across different administrative domains and physical locations but still keeping the actual data local (without unnecessary replicated data storage or transmission). This aligns with

2.4 Trusted Framework: RE4DY

RE4DY provides a toolkit that enables the deployment, operation, and management of distributed data ecosystems where different stakeholders can automatically share, consume, and integrate data across value networks. The RE4DY toolkit enables the standardization of data assets, provides the means to ensure data quality and compliance when accessing the data, and provides support for data sharing and integration. The use of IDS-certified components [1], in alignment with the data space specifications defined by IDSA [7] and Gaia-X [3], enables data sovereignty and secure data sharing based on specific data contracts and fosters trust among participants. More concretely, the main components included in the toolkit that enable the participation in the Data Space are a Dynamic Attribute Provisioning Service (DAPS), which allows the identification of the participants, and IDS connectors, which mediate the communications among the different stakeholders in the data space. Hence, the RE4DY toolkit simplifies and accelerates the process to achieve trusted data sharing among stakeholders and increases the use and reuse of data assets. Finally, it should be noted that the toolkit also facilitates data connectivity beyond the scope of the RE4DY project, thus increasing the potential for data asset integration and reuse.

2.5 Data Spaces: Sovity

The Sovity Connector [10] is a lightweight, standards-compliant implementation of the IDS Connector specification [5] that facilitates participation in data spaces. Developed to align with the architectures promoted by IDSA and Gaia-X, the Sovity Connector enables organizations to share and consume data in a trusted, sovereign and policy-compliant manner. Key features of the Sovity Connector include the following:

- Policy enforcement: Supports the use of usage control mechanisms via the Open Digital Rights Language (ODRL) [8] or similar frameworks, ensuring that data usage adheres to the provider's terms and conditions.
- Identity and trust: Leverages Self-Sovereign Identity (SSI) and Decentralized Identifiers (DIDs) to authenticate and authorize participants in a federated environment, reducing dependency on centralized identity providers.
- Interoperability: Fully compatible with IDS Reference Architecture [6] and Gaia-X federation services [4], ensuring seamless integration in multi-vendor, cross-sectoral ecosystems.
- Plug-and-play deployment: Designed to be easy to install and configure, the Sovity Connector reduces entry barriers for small and medium enterprises (SMEs) and public administrations to join data spaces.

From a technical perspective, the Sovity Connector is built using modern microservice principles, offering APIs for data catalog publishing, contract negotiation, and secure data transmission. It abstracts the complexities of the underlying IDS protocol stack, enabling organizations to focus on business logic and data governance rather than implementation specifics.

3 Proposed Solution

3.1 Overview of the Framework

The proposed framework (Fig. 1) integrates the aerOS Data Fabric with the RE4DY Data Space to allow access to the Data Products in a trusted environment with fine-grained access control policies. The aerOS Data Fabric defines a federated environment to connect different stakeholders who can define Data Products that can be discovered and accessed within the Data Fabric. Moreover, the Data Fabric can integrate resources deployed from Edge to Cloud. The integration of the aerOS Data Fabric with the Data Space allows the definition of fine-grained access control policies to share selected Data Products with third-parties in a trusted environment and enables the monetization of those assets. Thus, the main components of the proposed solution complement each other.

In the proposed framework, a data provider can create a Data Product in the aerOS Data Fabric, register it in the Data Space catalog, and define the proper data access policies. The metadata of the data asset can then be found in the data catalog, making the data asset discoverable. A data consumer from the Data Space who is interested in using the data can negotiate and obtain access based on the policies defined by the data provider (Fig. 1).

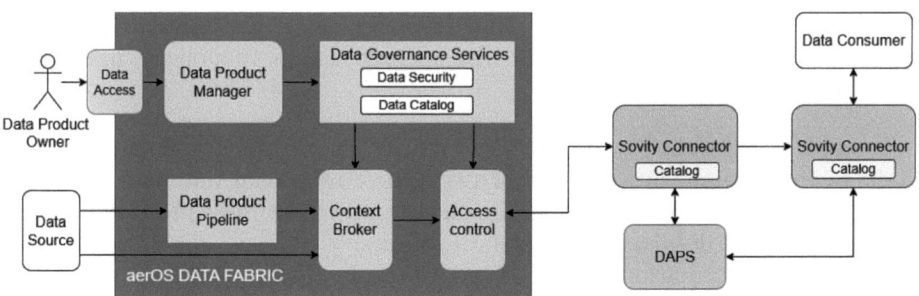

Fig. 1. Overview of the integrated framework.

3.2 Technological Implementation

Data Products. The Data Product Manager component takes the form of a containerized REST API server with orchestration capabilities in the backend. The Data Product Manager has been developed in Python, leveraging the

FastAPI library. Building upon the proposed definition of a data product, the REST API expects the following metadata and artifacts for onboarding new data products in the Data Fabric:

- Data Product Creation:
    - Data source configuration: Indicates the type of data source along with connection details such as source URL (e.g., JDBC URL in relational databases) and access credentials (e.g., username/password, certificate). Additionally, the following information might be provided depending on the type of data source: (1) Data source freshness: Only supported for data sources of batch type. Determines how frequently Data Fabric collects raw data from the target data source. (2) Data product serving: (A) Materialization (by default), which stores the data in the Context Broker; (B) Virtualization, which builds and serves the data product on-demand by means of a Context Source.
    - Mapping, which can be (1) a file with declarative rules (e.g., TRL) describing the mappings to transform ingested raw data into a graph, and semantically annotate against an ontology, and (2) programmatic, which relies on a custom application developed ad-hoc for such mapping.
- Data Governance:
    - Governance metadata: Identifiers to entities required for governing the new data product from the data catalogue. These identifiers are used by the respective entities in the graph: Data domain, Data product owner, Data product developer, Business glossary terms, and Tags/keywords.
    - Access control policies: A typical scenario is that of a data product made available in the catalogue, but to which no-one has access. Consumers need to search for the data product and to send a request to the data owner to access it (establishing a kind of data contract with the purpose, data consumer identification and grant access for limited time). Furthermore, the data owner could also include access to policies upon data product onboarding.

In aerOS Data Fabric, the Data Product Manager supports onboarding data products from batch data sources of two types: relational database and files. Figure 2 includes snapshots of the documentation of the data product onboard API for relational database and file data sources.

Taking the information provided by data product owners through the REST API, the Data Product Manager deploys and configures a new data pipeline for the onboarded data product. In this regard, all components of the Data Product Pipeline are containerized and can be deployed as Helm Charts on Kubernetes. To orchestrate the construction of the pipeline in IEs that belong to K8s clusters, the Data Product Manager leverages the Helm Controller component from the FluxCD project. The Helm Controller implements a Kubernetes operator that defines Helm charts and Helm releases as new custom resources in Kubernetes. This allows managing the life cycle of Helm releases through the K8s API.

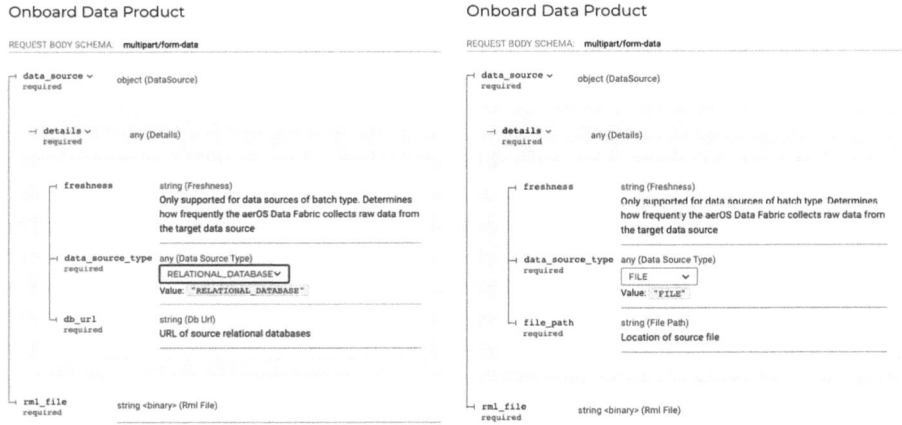

Fig. 2. Data product onboard API.

Data Catalogue. The Data Catalogue organizes the data available in the Data Fabric and makes the data assets discoverable. It can be integrated with LDAP via the LDAP connector, which collects data from LDAP-compliant databases, like the open-source project OpenLDAP, and transforms the LDAP data into NGSI-LD data that are eventually persisted in the Orion-LD Context Broker (Fig. 3). The LDAP Connector has been implemented as a containerized Python application that leverages the ldap3 library to interact with OpenLDAP, and the python-ngsi-ld-client to interact with the Orion-LD Context Broker.

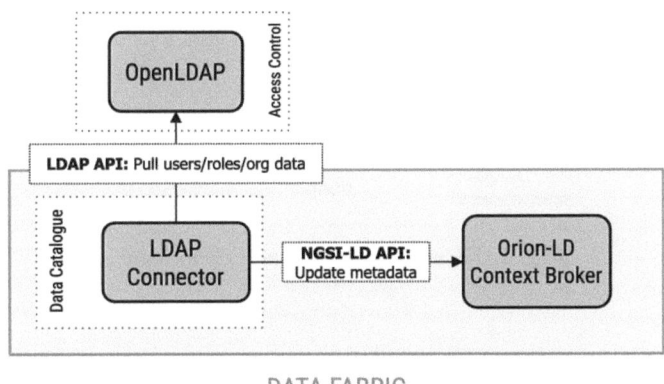

Fig. 3. LDAP integration.

During the workflow, the LDAP Connector periodically syncs with the LDAP database and maps the data to concepts of standard ontologies PRO and ORG, such as Person (User), Role, and Organization. These mappings are currently under exploration, and further details will be provided in the next release of the aerOS Data Fabric.

Federation to Achieve Domain Distribution. The aerOS Federator serves as a management service responsible for controlling the establishment and maintenance of federation mechanisms among the multiple aerOS domains that form the continuum and host Data Fabrics. It is composed of two main components (Fig. 4): (I) custom aerOS Federator component and (II) Orion-LD context broker. The core federation functionalities are provided by the context broker through the establishment of Context Source Registrations (CSR), which allows an Orion-LD instance to retrieve information (in NGSI-LD entities format) from another Orion-LD instances, which in the aerOS continuum means that data from one domain can be obtained just by calling the context broker of another domain once a proper registration has been performed.

This capability is directly linked with the aerOS Data Fabric described above, adding the crucial feature of secure and resilient federation, designed to act as the starting point of the mechanism (domain discovery), which leverages Context Source Registrations to achieve domains federation, always having into consideration the ontologies of the Data Products defined in the Data Fabric.

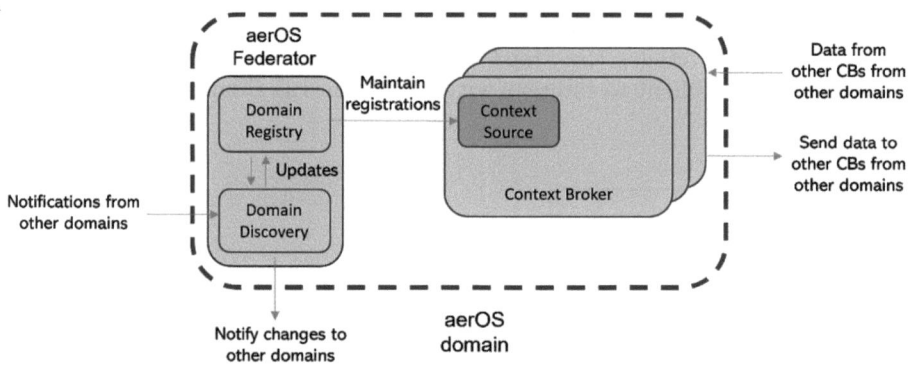

Fig. 4. Federation among aerOS domains.

Data Space Connector. In a federated data space, the Sovity Data Space Connector, which is based on the Eclipse Dataspace Components (EDC) [2], provides REST APIs for publishing data assets, defining usage contracts, and exchanging data with policy enforcement. This connector is implemented in Java and it is typically deployed as a set of microservices.

The aerOS Data Fabric can integrate with a Sovity Connector to register aerOS data products in the shared Data Space and to consume assets from other participants. Here, we outline key Sovity API endpoints and show how aerOS

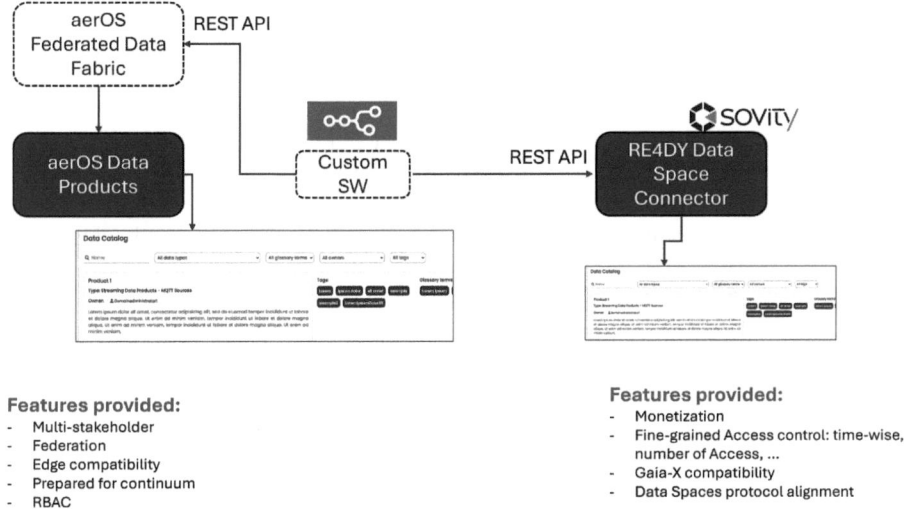

Fig. 5. Overall integrated framework.

Data Fabric components can invoke them in common workflows (such as asset publication, access request/negotiation, and data delivery), with example JSON payloads. We also discuss the monetization schemes and identity/access-control setup (e.g. OAuth2 or SSI) that binds the two systems.

The Sovity Connector implements the standard EDC control-plane APIs for assets, policies, and contracts, plus enhanced "wrapper" endpoints for streamlined UI or use-case interactions. Key endpoint categories include the following:

- Asset and Catalog Management: Registering a data asset (catalog entry), optionally attaching metadata and a data source (data address). Assets become searchable catalog entries.
- Policy Definitions: Defining usage policies (based on ODRL and defined in the API call as constraints in JSON). Policies (permissions, prohibitions, obligations) are referenced by contract definitions.
- Contract Definitions: Binding assets to policies through contract offerings.
- Catalog Queries: Browsing offers. Consumers retrieve data offers from peers.
- Contract Negotiation: The consumer initiates negotiation via the EDC control API. The request must include the target connector's ID or address, the asset ID or contractDefinition ID being requested, and the usage policy (often in JSON-LD ODRL form). The connector then returns a negotiation or agreement ID once terms are finalized.
- Data Transfer (Artifacts): After agreement, the consumer starts data transfer. The payload specifies the contractAgreementId, transfer type (e.g., HTTP push or pull), and a data sink address (endpoint where data should be delivered). The connector will then enforce the contract and transmit the data.
- Identity and Access: Sovity connectors rely on an external Identity Provider. In practice, the connector's API is secured by OAuth2. There is no user-

lookup endpoint per se, but connectors expose token endpoints to exchange client credentials or certificates for access tokens. Connectors are validated via either a DAPS or a SSI hub.

Together, these APIs would let aerOS Data Fabric components publish and use data offers programmatically: assets and contracts are created via the management API (or Sovity's bundled wrapper), and data flows occur through the data-plane endpoints once contracts are in place. Sovity emphasizes data sovereignty: "Connectors enable peer-to-peer data exchange while maintaining data sovereignty" by enforcing usage policies.

The Sovity Connector supports data monetization through contractual mechanisms and policy enforcement. When integrated with the aerOS Data Fabric, Sovity enables usage-based, subscription-based, or flat-fee data monetization models within federated data spaces. Sovity enables dataset monetization using commercial contract offers that define payment terms, pricing models, and obligations. The key monetization models are: i) Pay-per-Access / One-Time Purchase. where the consumer pays once for a single data transfer (e.g., download of a historical dataset); ii) Subscription-Based Access, where the consumer pays a recurring fee for continuous access (e.g., access to a real-time sensor feed or regularly updated weather data); and iii) Usage-Based Billing (Metered), where payment is based on actual usage (e.g., per API call, per MB, per transaction); this model could be used in on-demand traffic or satellite imagery services.

End-to-End Integration. Secure interaction requires mutual identity verification and credential handling. Sovity connectors authenticate to each other via OAuth2 tokens (from a DAPS) or via SSI (Managed Identity Wallets). Linking with aerOS Data Fabric approach, Keycloak would be leveraged implementing OAuth2 tokens and access control to HTTP API endpoints.

In practice, the Sovity Connector can be configured to trust tokens issued by the aerOS Identity Manager (IdM) (e.g. via OpenID Connect or a Keycloak instance deployed for that purpose). Data Fabric clients (or the DPM) obtain an access token from their IdM and include it in API calls to Sovity Connector (e.g., in the Authorization header when calling the management or wrapper APIs). Sovity's connector verifies this token (via its own Keycloak instance or a trust broker). Conversely, when the Sovity connector initiates callbacks or transfers, it can also use an OAuth2 token (its own client credentials) which aerOS must trust. In effect, both sides rely on a shared trust anchor or token exchange.

Importantly, both systems operate under the principle that requester and provider establish trust without sharing raw credentials. aerOS' OAuth2 federation and Sovity's DAPS can be bridged: for instance, aerOS tokens could be accepted by Sovity's DAPS as proof of identity, or Sovity's identity provider could be integrated as an external IdP to aerOS.

The first step to share data using the proposed framework is the definition and publication of the data asset. The process is defined as follows:

1. Define Data Product in aerOS: A user or process creates a Data Product in aerOS, which stores the associated metadata (e.g. IDs, schema, catalog info) in a federated approach.
2. Create Sovity Asset: The DPM calls the Sovity Connector to register the asset. For example, using Sovity's UI wrapper (Listing 1.1), aerOS defines a paid asset ("asset-123") that requires a one-time €500 payment before transfer and with a restricted policy (only use for "analytics"). On success, Sovity returns an ID (same as "asset-123").
3. Result: The asset appears in the Sovity Connector's data catalog, visible to peers in the federated space. The corresponding ContractDefinition linking this asset to the new policy is registered in the connector. Sovity ensures that the metadata is stored and shared in its local/federated catalog.

Listing 1.1. Example of asset registration using Sovity's UI wrapper

```
POST /wrapper/ui/pages/create-data-offer
Content-Type: application/json

{
  "asset": {
    "id": "asset-123",
    "title": "High-Res Satellite Imagery",
    "description": "High-Res Satellite Imagery",
    "publisherHomepage": "https://aeros.example.org",
    "dataSource": {
      "type": "HTTP_DATA",
      "baseUrl": "https://dpm.aeros.example.org/data/orderdb",
      "method": "GET"
    }
  },
  "publishType": "PUBLISH_RESTRICTED",
  "policyExpression": {
    "type": "AND",
    "expressions": [
      {
        "type": "CONSTRAINT",
        "constraint": {
          "left": "dataspace:purpose",
          "operator": "EQ",
          "right": {"type": "VALUE", "value": "analytics"}
        }
      },
      {
        "type": "CONSTRAINT",
        "constraint": {
          "left": "edc:businessTerm",
          "operator": "EQ",
          "right": {
            "type": "VALUE",
```

```
                "value": "PAY_BEFORE_USE"
              }
            }
          }
        ]
    },
    "pricing": {
      "price": 500,
      "currency": "EUR",
      "billingModel": "ONE_TIME"
    }
}
```

Once the data product has been created and registered, data consumers can discover it and negotiate access as follows:

1. Discovery: A consumer queries the catalogs (Listing 1.2). Sovity returns offers including "asset-123" and its terms.
2. Negotiate Contract: The consumer initiates a contract negotiation on Sovity's API. An example payload is shown in Listing 1.3, where *counterPartyId/address* identifies the provider connector, *contractOfferId* refers to the provider's published contractDefinition, *assetId* is the asset being requested, *policyJsonLd* restates the intended usage and pricing policies (which could mirror the published one), and *callbackAddresses* indicates Sovity where to POST updates. Sovity processes this request via IDS protocols. If the policies are compatible, it will establish a ContractAgreement, returning an ID.
3. Agreement: Once the connector finalizes the negotiation, it issues a contract agreement ID (e.g. "contract-agreement-789"). At this point, both aerOS systems have a binding contract: the consumer may consume the data and the provider must deliver it under the stated usage policy.

Listing 1.2. Example of query to the catalog

```
GET /wrapper/ui/pages/catalog-page/data-offers?
connectorEndpoint=https://provider-connector.aeros.example.org
&participantId=provider-org
```

Listing 1.3. Example of contract negotiation

```
POST /wrapper/ui/pages/catalog-page/contract-negotiations
Content-Type: application/json

{
  "counterPartyId": "provider-connector-001",
  "counterPartyAddress":
    "https://provider-connector.aeros.example.org/api/ids/data",
  "contractOfferId": "contractdef-asset-123",
  "assetId": "asset-123",
  "policyJsonLd": "{ \"@type\": \"ContractPolicy\",
    \"permissions\": [{\"target\":\"asset-123\",
```

```
      \"action\":{\"type\":\"USE\"}}] }",
  "callbackAddresses": [
    {
      "uri": "https://consumer-app.aeros.example.org/notify",
      "events": ["CONTRACT_NEGOTIATION_FINALIZED",
      "TRANSFER_PROCESS_COMPLETED"],
      "authHeaderName": "Authorization",
      "authHeaderVaultSecretName": "ConsumerApiKey"
    }
  ]
}
```

Finally, if the negotiation is successful, the data consumer gets access to the data under the defined conditions. The data transfer process has the following steps:

1. Initiate Transfer: With an agreement in hand, the consumer triggers data transfer. For example, using Sovity Connector's transfer API (Listing 1.4), where *contractAgreementId* identifies the contract, *"type":"HTTP_DATA_PUSH"* tells Sovity Connector to push the data to the consumer, and *httpDataPush* contains the consumer's data receiver endpoint and any needed auth (e.g., an OAuth bearer). Sovity Connector will perform an HTTP POST of the data payload to that URL, enforcing the contract.
2. Data Flow: The Sovity Connector (provider side) fetches the data from the aerOS data source and then sends it to the consumer's URL. It logs the transfer and enforces any policies (e.g., it may check that the OAuth token presented is valid). Once transfer is complete, Sovity can notify both parties via the callback (if configured).
3. The data has been delivered securely. The aerOS consumer can now process it. The Sovity Connector records the transfer in its history (accessible through the UI or API), which aerOS could query if needed.

These sequences assume that both sides have established trust. In each step, Sovity Connector's APIs enforce the agreed policies: e.g., if a policy forbids copying, the Sovity Connector will not let the consumer "download" the asset via any other channel.

Listing 1.4. Example of data transfer request

```
POST
/wrapper/ui/pages/contract-agreement-page/initiate-transfer-v2
Content-Type: application/json

{
  "contractAgreementId": "contract-agreement-789",
  "type": "HTTP_DATA_PUSH",
  "httpDataPush": {
    "baseUrl":
      "https://consumer-app.aeros.example.org/api/data-receiver",
    "method": "POST",
```

```
    "headers": {
      "Authorization": "Bearer <token-from-consumer>"
    }
  }
}
```

Table 1 provides a summary of the integration workflow between the Sovity Connector and the aerOS Data Fabric, omitting the payment-specific steps.

Table 1. Sovity Connector workflow summary

Step	Component	Action	Details
1	aerOS DPM	Define Data Product	Configure metadata for the dataset (ID, title, description, access URL, usage model).
2	DPM to Sovity Connector	Publish Data Offer	Use Sovity's wrapper API to register the dataset as an asset and create a corresponding contract definition and usage policy.
3	Sovity Connector	Expose in Catalog	Asset becomes visible in the Sovity Connector catalog with usage conditions (e.g., restricted, read-only, etc.).
4	Consumer	Browse Catalog	Consumer discovers available data offers via Sovity Connector's catalog API or UI.
5	Consumer to Sovity Connector	Initiate Contract Negotiation	Consumer selects the offer and sends a contract request via Sovity's negotiation API.
6	Sovity Connector	Enforce Usage Policy	Connector evaluates the contract request against the associated policy (e.g., usage constraints, obligations).
7	Sovity Connector	Sign Contract	If policy conditions are met, a formal contract agreement is established.
8	Sovity Connector	Initiate Data Transfer	Data is transferred according to the contract terms using a supported protocol (e.g., HTTPS, S3, etc.).
9	aerOS (optional)	Monitor Access & Usage	aerOS can observe usage logs and enforce additional governance or compliance checks.

4 Conclusion and Future Work

This paper describes how the Federated Data Fabric developed in the aerOS project can be integrated with a Data Space to enable trusted fine-grained access to selected data by third parties. The aerOS Data Fabric provides a unified interface to access the data from the IoT-Edge-Cloud continuum and integrates data governance and traceability. Moreover, it supports federation across different domains.

The integration presented in this paper allows data product owners to share data from the continuum in multi-vendor cross-sectorial data ecosystems. This integration facilitates the creation of public or commercial datasets that can be easily discovered and used by other parties in the data space under predefined terms, thus supporting the creation of new data-driven solutions. Moreover, data from other participants can be accessed by the Data Fabric and potentially enrich the available data assets. This collaboration could facilitate the reuse and monetization of data and result in faster innovation. Furthermore, since the integrated framework described in this paper aligns with European Data Initiatives and standards, it would facilitate data integration and decentralized data management across different sectors while maintaining data sovereignty.

Future work will focus on the validation of the integrated framework in scenarios based on real-word use cases.

Acknowledgments. This research has been funded by the European Commission, under the Horizon Europe projects aerOS (grant agreement 101069732) and RE4DY (grant agreement 101058384).

Disclosure of Interests. The authors have no competing interests to declare that are relevant to the content of this article.

References

1. Certification - IDSA. https://internationaldataspaces.org/offers/certification/
2. Eclipse dataspace components. https://projects.eclipse.org/projects/technology.edc
3. Gaia-X: A federated secure data infrastructure. https://gaia-x.eu/. Accessed 30 June 2025
4. GXFS and the XFSC toolbox - GXFS.eu. https://www.gxfs.eu/set-of-services/
5. Ids connector-ids knowledge base. https://docs.internationaldataspaces.org/ids-knowledgebase/ids-ram-4/layers-of-the-reference-architecture-model/3-layers-of-the-reference-architecture-model/3_5_0_system_layer/3_5_2_ids_connector
6. Ids reference architecture model 4.0. https://docs.internationaldataspaces.org/ids-knowledgebase/ids-ram-4
7. International data spaces. https://internationaldataspaces.org/
8. ODRL information model 2.2. https://www.w3.org/TR/odrl-model/
9. Re4dy | manufacturing data networks. https://re4dy.eu/. Accessed 30 June 2025
10. sovity: Sovereign data exchange with your partners in data spaces. https://sovity.de/en/sovity-en/
11. aerOS project: Autonomous, scalable, trustworthy, intelligent European meta Operating System for the IoT edge–cloud continuum. https://aeros-project.eu/ (2025). Accessed 30 June 2025
12. aerOS Consortium: D2.6 – Aeros architecture definition (1). Report, aerOS Consortium (2024). https://aeros-project.eu/wp-content/uploads/2024/09/aerOS_D2.6_1.pdf. Accessed 30 June 2025
13. Cuñat, S., Julian, M., Belsa, A., Valero, C.I., Esteve, M., Palau, C.E.: Secure, Trusted, Privacy-Protected Data Exchange in an Edge-Cloud Continuum Environment, pp. 201–231. Springer Nature Switzerland (2024). https://doi.org/10.1007/978-3-031-58388-9_7
14. Martinez-Casanueva, I.D., Bellido, L., González-Sánchez, D., Lopez, D.: CANDIL: a federated data fabric for network analytics. Futur. Gener. Comput. Syst. **158**, 98–109 (2024)
15. Nagel, L., Lycklama, D.: Design principles for data spaces - position paper (2021). https://doi.org/10.5281/zenodo.5105744

16. Vaño, R., Lacalle, I., Palau, C.E.: Federation of distributed domains in the cloud-edge-IoT continuum. In: Proceedings of the 2nd International Workshop on MetaOS for the Cloud-Edge-IoT Continuum, pp. 14–19. MECC '25, Association for Computing Machinery, New York, NY, USA (2025). https://doi.org/10.1145/3721889.3721923

Open Access This chapter is licensed under the terms of the Creative Commons Attribution 4.0 International License (http://creativecommons.org/licenses/by/4.0/), which permits use, sharing, adaptation, distribution and reproduction in any medium or format, as long as you give appropriate credit to the original author(s) and the source, provide a link to the Creative Commons license and indicate if changes were made.

The images or other third party material in this chapter are included in the chapter's Creative Commons license, unless indicated otherwise in a credit line to the material. If material is not included in the chapter's Creative Commons license and your intended use is not permitted by statutory regulation or exceeds the permitted use, you will need to obtain permission directly from the copyright holder.

Data Spaces and Digital Infrastructure for the IoT Era

Web Based Monitoring, Orchestration and Simulation

Dave Raggett$^{(\boxtimes)}$

ERCIM/W3C, Biot, France
dave.raggett@ercim.eu

Abstract. This paper describes a framework for web-based monitoring, orchestration and simulation for industrial settings such as highly automated factories and warehouses. HTML5 is used for a 2.5D visualization with local prediction for smooth animation and 3D models for robot arms. The web server aggregates data from the devices for streaming to the web page, for generating situation reports, and enabling users to intervene as needed. Simulation is possible using simple looping sequences of actions. A cognitive architecture is described for richer behaviour that dynamically adapts to the context, decoupling reasoning from real-time control using asynchronous intents. Stochastic rules allow agents to escape from futile patterns of behaviour. Agents can control multiple devices. For a larger numbers of devices, control can be distributed across multiple agents in a way that decouples applications from the underlying protocols and addressing schemes. Iterative refinement is applied as new requirements come to light, e.g. to handle faults, or where humans have intervened in unexpected ways. The paper closes with a short summary of previous work and suggestions for future work on integrating generative AI to further simplify development and provide supervisory control.

Keywords: automation · cognitive control · swarm computing · orchestration · simulation · multi-agent systems

1 Introduction

In the context of control systems, agents are computer systems that can operate autonomously, making decisions and taking actions without constant human oversight. Agent behaviour can be classified into broad categories, e.g., reactive, model-based, goal-based, and utility based, as well as on how they interact with other agents, e.g. collaborative or competitive. Reactive agents take actions purely on the basis of current information on their environment. Model-based agents build and exploit models of their environment, e.g. a robot vacuum cleaner that builds a map of the rooms it operates in. Goal-based agents combine models with search and planning. Utility-based agents have a function for scoring the desirability of different states or outcomes, and focus on maximising the utility, not just fulfilling a goal.

Building upon the capabilities of such agents, this paper describes approaches to web-based monitoring, orchestration and simulation for industrial systems. Specifically,

© The Author(s) 2026
M. Presser et al. (Eds.): GIECS 2025, CCIS 2719, pp. 111–124, 2026.
https://doi.org/10.1007/978-3-032-09555-8_7

web-based monitoring leverages web browsers to observe the real-time state of industrial environments like factories or warehouses. This monitoring functionality can then be extended to facilitate orchestration, providing a comprehensive means for coordinating, synchronizing, and overseeing complex, automated workflows across diverse systems and applications.

Orchestration can be applied to either control an industrial system or to drive a simulation of an industrial system. This paper describes a basic approach to describing repeated fixed behaviours that are typical of current industrial control systems, and the extension to more flexible context sensitive adaptive rule-based control, and the means to synchronise behaviour in a multi-agent system. Rule-based orchestration uses rules to define the order, conditions, and dependencies of these behaviours, ensuring they operate together smoothly.

The paper concludes with an assessment of opportunities for further work, e.g. on decentralised decision making, multi-agent reinforcement learning, and the combination of generative AI with rule-based approaches.

2 Web-Based Monitoring

Users can use the web browser to see a 2.5D isometric view of the factory or warehouse floor, with the means to pan and zoom the view using game controllers, keystrokes or gestures on the computer's touchpad. Isometric views show objects at the same size regardless of their location, just as is the case for 2D maps. Users can further interact with the presentation to request further information and to change system parameters.

Figure 1 shows a screenshot from a simulation of a smart warehouse [2] where robot forklifts move pallets carrying different kinds of products from the incoming trucks (yellow) to the outgoing trucks (red) to fulfil customer orders. Pallets are temporarily stored in the racks if appropriate. Forklifts are assigned jobs, and plan the shortest route following the traffic lanes (marked in red and green) near the racks, and direct routes away from the racks. The forklifts are sent to recharge when their battery level falls below a given threshold. Each forklift uses its first-person view to avoid other forklifts, and all forklifts stop to avoid colliding with nearby humans. The user can request information on a given forklift, see the pop-up pane for an example.

3 Digital Twins

In this paper, we define digital twin as a term for a computer model of physical systems and processes. Digital twins can be used for:

- monitoring current state of their physical twins
- controlling their physical twins
- diagnosing and fixing faults using causal models
- planning and simulation
- optimisation based upon applying machine learning to recorded data

This paper focuses on web-based monitoring, orchestration and simulation. The starting point is a web server that hosts the digital twin models as virtual objects with properties, actions and events, as inspired by the Nephele project [3] and the W3C Web of Things [4].

For monitoring, we need to dynamically update the state of the digital twins from the physical systems and processes, and to stream the updates to the web pages connected to the web server. Orchestration is the process for remotely managing industrial systems. It can apply to controlling the actual devices, or it can be used in a simulation to control virtual digital twins.

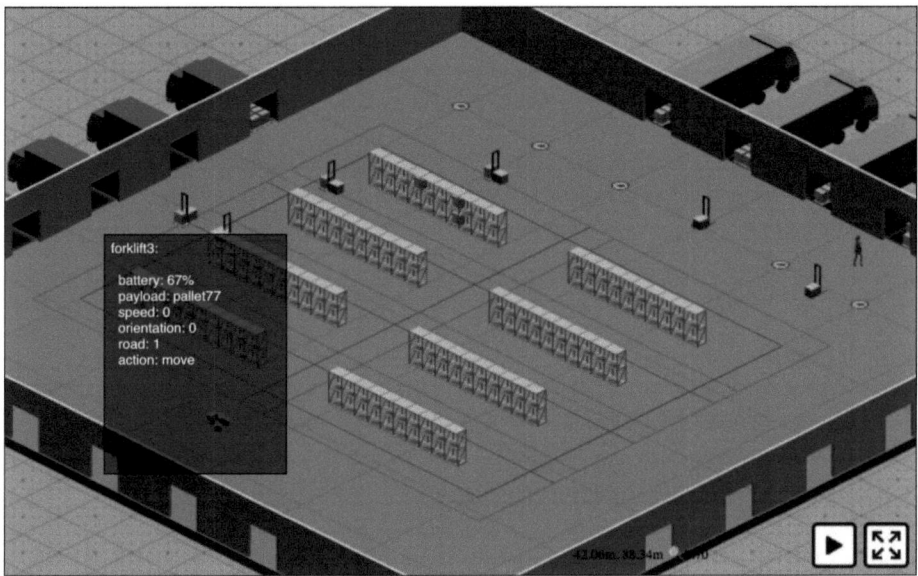

Fig. 1. Screenshot from SimSwarm, a simulation of a smart warehouse operated by robot forklifts. See the demo at: https://www.w3.org/Data/demos/chunks/warehouse/.

The web server acts as a relay, communicating with the devices using whatever protocols are appropriate, e.g. HTTP, MQTT or DDS, to update the digital twins held by the server.

The author has developed an opensource NodeJS solution that combines a JavaScript module for HTTP + Web Sockets with a custom module that updates the digital twin models, and streams the updates as JSON messages to the web pages currently connected to the server. NodeJS is a strong community with plenty of support for a wide range of protocols [5].

If your devices feature ROS (the robot operating system), then you will find it convenient to use the *rosnodejs* module [6] for server-side communication with a networked ROS Master. Alternatively, you can write your own drivers, e.g. using one of several DDS modules for NodeJS with a mapping between IDL and JavaScript datatypes (Figs. 2 and 3).

```
let pallet1 = new Pallet({
    name:"pallet1",
    x:20,
    y:30,
    orientation:0,
    loaded:true
});
let forklift1 = new Forklift({
    name:"forklift1",
    x:10,
    y:30,
    z:0,
    orientation:0,
    speed:1,
    forkspeed:0.5,
});
```

Fig. 2. Example for declaring a pallet and a forklift.

```
# example message for adding a pallet with ID "pallet2"
{"name":"pallet2","type":"pallet","x":10,"y":40,
    "orientation":3,"loaded":false}
# example message for updating forklift state
{"name":"forklift1","type":"forklift","x":16.58,"y":30,"z":0,
    "orientation":0,"speed":1,"held":false}
```

Fig. 3. Examples of JSON messages sent by the server to update connected web pages.

4 Simple Sequences of Instructions

In today's factories it is commonplace to program robots and other factory machinery to indefinitely repeat the same sequence of instructions. The JavaScript module for digital twins provides a simple means to provide such instructions:

The *move* action prepends a *turn* action to turn to the appropriate *orientation*. Likewise, the *grab* action prepends *move*, *turn* and *forks* actions as needed, having selected which side of the pallet is the nearest to grab the pallet. To raise or lower the forks explicitly use the *forks* action, e.g. {act:"forks", z:0.5} which raises or lowers the forks to 0.5m above the floor. To turn to a given orientation, you can use the *turn* action, e.g. {act:"turn", orientation:4}, noting that orientation is an integer in the range 0 to 7 reflecting the eight camera orientations for the 2.5D image tiles used in rendering (Fig. 4).

```
forklift1.setTask([
    {act:"wait", time:4},
    {act:"grab", pallet:"pallet1"},
    {act:"move", x:10, y:30},
    {act:"release"},
    {act:"move", x:20, y:30},
    {act:"grab", pallet:"pallet1"},
    {act:"move", x:20, y:30},
    {act:"release"},
    {act:"move", x:10, y:30},
    {act:"next"}
]);
```

Fig. 4. Example declaring a sequence of control instructions for a forklift.

You can instruct the forklift to wait for a fixed number of seconds, e.g. {act:"wait", time:4} waits for 4 s. You can make it a little more interesting by instructing a wait that is randomly selected between a minimum and maximum duration, e.g. {act:"wait", min:1, max:5} waits between 1 and 5 s.

The *next* action defaults to transferring control to the first action in the task. However, if you provide a list of named steps, one of those steps is randomly selected.

Note that the *move* action doesn't itself include support for avoiding obstacles, whether stationary or moving. This is something to consider in future work on integrating Chunks & Rules for more flexible behaviour, and as part of a plan for scaling up to a much larger range of factory devices, e.g. robot arms, conveyor belts, palletization and manufacturing cells. The SimSwarm demo uses collision avoidance heuristics implemented directly in JavaScript. This could be replaced by an explicit system of rules.

5 Cognitive Control for Context Sensitivity

Simple repeating sequences of instructions are fine most of the time, but can run into difficulties in the case of faults or when a human has intervened in some manner that upsets the assumptions in the programming. An example is where a robot tries grab something which isn't in quite the correct position. Another case is where a human worker has removed something that the robot expects to find.

5.1 Chunks and Rules

This is a syntax and open-source implementation for facts and rules inspired by John Anderson's cognitive architecture ACT-R, which is grounded upon decades of research in the Cognitive Sciences [7]. It is at a higher level than RDF and uses a simple easy to use syntax.

The architecture features one or more cognitive modules, each of which is associated with a buffer that can contain just one chunk. A chunk is a set of properties including a type and a unique chunk identifier. Property values are names, numbers or a list thereof. In respect to RDF, a chunk corresponds to a set of triples with the same subject node, and to RDF lists for list values. A convenient short hand is provided for unannotated relations (corresponding to RDF triples), which are internally reified into single chunks.

For more details, see the specification from the W3C Cognitive AI Community Group [8].

The architecture is motivated by the structure of the brain. The cognitive modules correspond to cortical regions. The rule engine corresponds to the basal ganglia in the centre of the brain. Cognitive buffers correspond to bundles of nerve fibres that connect the basal ganglia to the cortex. Each chunk thus corresponds to a vector in a high dimensional space (Fig. 5).

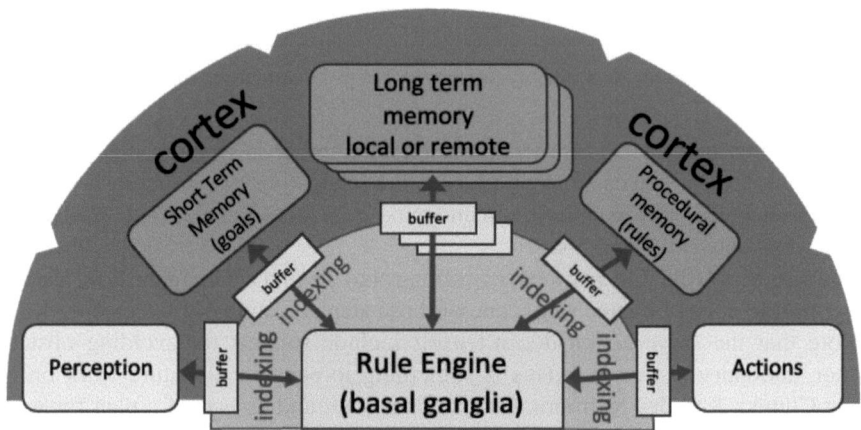

Fig. 5. Cognitive architecture for Chunks & Rules.

Rule antecedents are a conjunction of conditions expressed as chunks. Rule consequents are an unordered list of actions expressed as chunks. Rules are matched to the current state of the module buffers. If multiple rules are matched, a stochastic choice is made influenced by rule strength. Rule execution is sequential.

The approach borrows from ACT-R in respect to sub-symbolic parameters corresponding to the strength of facts and rules, mimicking the forgetting curve for human memory, likewise for spreading activation and the spacing effect. Performance is fast as a) the rules are matched to a handful of buffers rather than the entire set of chunk databases, and b) actions are asynchronous, enabling cognition to continue whilst actions are being executed. There is a suite of built-in actions for operations on chunk databases and an API for programming complex operations. Applications can register additional actions as needed, e.g. to operate a robot arm.

Chunks & Rules are well suited to orchestrating devices via their digital twins using intent-based actions for concurrent threads of behaviour. Intents specify the desired outcome rather than the details of how to realise that outcome, something best suited for delegation to specialist systems. Work is underway on extending Chunks & Rules to support messaging and synchronization across agents. This is described in a later section of this paper (Fig. 6).

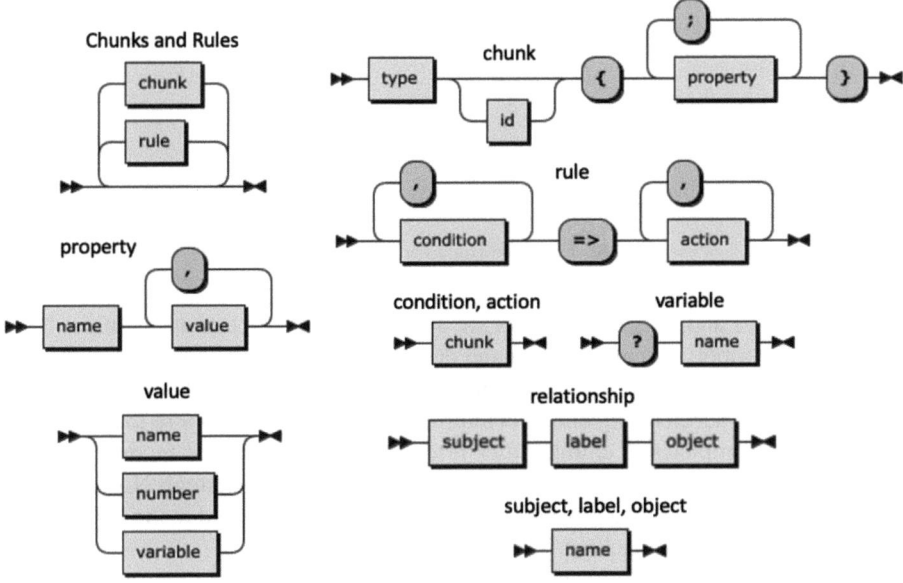

Fig. 6. Chunks & Rules syntax as railroad diagrams.

Note that you are free to use whitespace between tokens except between the "?" and name in a variable. Variables are used in rule conditions and actions (Fig. 7).

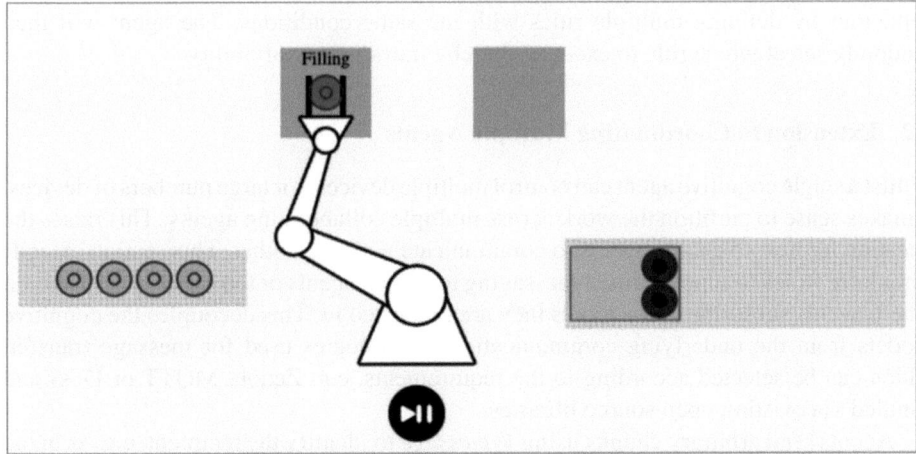

Fig. 7. Example web demo for bottling plant.

Here are some example rules from a web-based demo for a bottling plant [9] where a cognitive agent controls two conveyor belts, one robot arm as well as the filling and capping machines (Fig. 8).

```
# move robot arm into position to grasp empty bottle
after {step 1} =>
    robot {@do move; x -170; y -75; angle -180; gap 30; step 2}
# grasp bottle and move it to the filling station
after {step 2} =>
    goal {@do clear},
    robot {@do grasp},
    robot {@do move; x -80; y -240; angle -90; gap 30; step 3}
```

Fig. 8. Example rules for robot control.

Here the *move* and *grasp* commands are application defined and act as intent-based directives to the robot. The move command describes the position and orientation of the robot gripper along with the gap between its opposed fingers. This is expanded into a plan for smoothly accelerating and decelerating the various actuators in the robot arm. When the plan completes, a chunk "*after {step 3}*" is queued to the module buffer to trigger following behaviour. This demo uses some twenty rules in total. Note that rules can include variables which are instantiated when the rule is matched to the buffers. This is used in the demo when packing bottles six to a box.

The set of chunks and rules can be iteratively refined as new requirements come to light, e.g. in unusual situations like faults, or where humans have intervened in an unexpected way. This may involve the need for a systems programmer e.g. to update the machine vision system used to monitor physical processes, e.g. has a bottle fallen over on the conveyor belt? Has the bottle been correctly filled and capped?

In some circumstances, robots may find themselves trapped in futile repeating patterns of behaviour. To mitigate this, you can introduce stochastic non-deterministic behaviour by defining multiple rules with the same conditions. The agent will then randomly select which rule to execute, thereby introducing variability.

5.2 Extension to Coordinating Multiple Agents

Whilst a single cognitive agent can control multiple devices, for large numbers of devices, it makes sense to partition the work across multiple collaborating agents. This raises the question for how to enable agents to communicate with each other. This section extends Chunks & Rules to support direct messaging to named agents or topic-based distribution where agents subscribe to the topics they are interested in. This decouples the cognitive models from the underlying communication technologies used for message transfer, which can be selected according to the requirements, e.g. Zenoh, MQTT or DDS and handled via existing open source libraries.

Agents send arbitrary chunks using @*message* to identify the recipient, e.g. as in the following rule action property. You are free to set the chunk type and provide additional properties as appropriate to your needs. The addressee sees the chunk with the @message property stripped out. You can determine which agent sent the message by using @*src* with a variable in the conditions for a rule that matches the message (Fig. 9).

```
# tell agent4 to start
start {@message agent4}
```

Fig. 9. Example for direct messaging between agents.

Messages can be sent with *@topic* to subscribers of named topics, e.g. (Fig. 10)

```
#send stop message on topic12
stop {@topic topic12}
```

Fig. 10. Example for topic-based messaging between agents.

Agents can subscribe to topics with *@subscribe* as in the following example (Fig. 11):

```
# subscribe to topic12
listen {@subscribe topic12}
```

Fig. 11. Example for subscribing to a topic.

5.3 Agent Synchronisation Using Tasks

We can model a task as a sequence of rules that achieve some given aim. Each agent can be executing multiple tasks concurrently, at the same time as different agents are executing different tasks. To support task synchronisation, we need a means to make tasks explicit. This can be done using the *@do task* action property. This generates a unique task identifier. To access this identifier, use the *@task ?taskID* action property, which binds the variable to the new task identifier.

You indicate that a given task has succeeded or failed with the respective action properties *@do done* and *@do fail*, where you use *@task* to name which task you are referring to. The task identifier can be passed to the rule via a variable in one of the rule's condition chunks.

To synchronise across tasks, use *@all* to queue a chunk when all of the listed tasks have succeeded. Here is an example that initiates three tasks and then queues a chunk when they have all successfully completed. The action chunk with *@failed* is used to specify a chunk to queue when any of the listed tasks have failed (Fig. 12).

```
# rule to initiate several tasks and queue chunk
# process2 {} when all these tasks have completed

process1 { } =>
    a {@do task; @task ?task1},
    b {@do task; @task ?task2},
    c {@do task; @task ?task3},
    process2 {@all ?task1, ?task2, ?task3},
    recover2 {@failed ?task1, ?task2, ?task3}
```

Fig. 12. Example that starts three local tasks and initiates further behaviour when they have all succeeded or any one of them has failed.

Use *@any* in place of *@all* when you want to trigger behaviour when any one of the listed tasks have successfully completed. Note that you are free to set the chunk type and additional properties for action chunks with *@all, @any* and *@failed*.

@do task is used for tasks running on this agent. To initiate tasks on another agent you should use *@on*, as shown in the following example (Fig. 13):

```
# same thing, but this time on different agents
# initiating agent notified when tasks complete

process1 { } =>
    a {@do task; @on agent1; @task ?task1},
    b {@do task; @on agent2; @task ?task2},
    c {@do task; @on agent3; @task ?task3},
    process2 {@all ?task1, ?task2, ?task3}},
    recover2 {@failed ?task1, ?task2, ?task3}
```

Fig. 13. Example that starts three remote tasks and initiates further behaviour when they have all succeeded or any one of them has failed.

The *@on* action delegates a task to a named agent, and further ensures that the assignee informs the originating agent when the task succeeds or fails. Note that task identifiers are locally scoped to each agent, so that *@on* has the effect of running *@do task* on the assignee agent, generating a task identifier local to that agent. The reverse mapping of task identifiers is automatically applied when the assignee informs the assigning agent when a task completes.

5.4 Ideas for Further Exploration

Decentralised decision making has the potential for improved system resilience to faults and attacks. This can include the means for agents to take on and switch roles as needed, and the use of techniques based upon consensus, auctions and automated negotiation.

Agents may need to temporarily suspend tasks when they need to attend to higher priority interruptions *(@suspend ?taskID)*. When conditions permit, the task can be resumed *(@resume ?taskID)*. For this purpose, the task identifiers can be recorded in the chunk database for a given cognitive module and retrieved when needed. This is akin to asking yourself what was I doing before I was interrupted, and what should I do next?

It is natural to divide tasks across a hierarchy of sub-tasks. One way to achieve that is with *@all* and *@any*. This further relates to mechanisms to support reinforcement learning where the agent improves its skill by trying out different ways to achieve its goals. The agent needs to split its attention between working on getting better at a given task, and exploring fresh possibilities. Stochastic heuristics can be used to propose new rules to try out. We can then use success or failure to adjust the strengths of the rules to improve the agent's performance on the next attempt at this task.

One approach is to record the sequence of executed rules and propagate the reward backwards through time. From a biological perspective, this isn't convincing. Our episodic memory is better tuned to recalling events forwards in time rather than backwards. Another approach is to propagate rewards across the task hierarchy to update conditional probabilities for selecting a given rule in a particular context. This requires the agent to track the currently executing tasks and their relationships.

Single agent learning generalises to multi-agent reinforcement learning (MARL) where agents collaborate to explore the training space with the possibility of beneficial emergent behaviours. Agents can be fully cooperative, or more realistically, can have their own agenda with potentially conflicting goals.

As the number of agents is scaled up, inspired by human organisations, we are likely to require ways to limit agent to agent communication, e.g. via the environment (stigmergy), communication with physically nearby agents, or constrained by functional roles, using a scale-free peer-to-peer network where some agents are better connected than others. Agent to agent communication could be structured, e.g. JSON messages, or unstructured, e.g. using natural language, sound and images for agents based upon Generative AI.

Further inspiration comes from the relationship between the motion of individual water molecules and use of fluid mechanics to describe the motion of very large numbers of such molecules. If we can find efficient ways to describe the behaviour of large populations of agents, then we may be able relate the behaviour of an individual agent to the field equations for the population.

6 Longer Term Prospects

The success of Generative AI has shown the effectiveness of artificial neural networks for learning complex statistical relationships from large corpora. Recent work has focused on applying large language models to agentic frameworks. How does this relate to symbolic frameworks such as Chunks & Rules?

One idea is to focus on the generation of pertinent situation reports (SITREP). According to the Persimmon Group [10]: *The SITREP provides a clear, concise understanding of the situation, focusing on meaning or context in addition to the facts. It does not assume the reader can infer what is important; rather, it deliberately extracts and highlights the critical information. A good SITREP cuts through the noise to deliver exactly what matters: what is happening, what has been done, what will be done next, and what requires attention or decision.*

In principle, we could use large language models to generate SITREPs using plugins to access external systems in what is a neurosymbolic approach. An agentic framework such as *langchain* [11] could be used to guide a large language model (LLM) through engineering the prompts and interpreting the responses in an extended dialogue.

Another idea is to train LLMs to generate chunks and rules conditioned by natural language descriptions, so that users can describe their objectives rather than having to specify the details of how to fulfil those objectives. Generative AI is well suited to dealing with gaps and ambiguities in their input. The feasibility of this approach is boosted by success on accelerating conventional programming tasks. The main challenge is to curate a large enough corpus of training data. If this can be achieved, we could then use LLM based agents to supervise lower-level agents using symbolic approaches.

An intriguing research question is whether neural networks are better suited to controlling industrial systems than symbolic approaches? In principle, neural networks could be better suited for managing ad hoc situations involving ambiguity and novelty based upon their breadth of knowledge. Neural networks could likewise offer greater performance at learning new skills given back propagation and gradient descent as a means to train models with billions of parameters. This will be particularly effective when training models in risk-free simulated environments rather than having to try everything with real robots.

7 Relationship to Previous Work

In some cases, you want the control behaviour to always be exactly the same in the same context. In other cases, non-deterministic behaviour is appropriate when flexibility, adaptability, and the means to handle uncertainty are needed. They allow multiple possible outcomes and can incorporate randomness or probabilistic decision-making. Chunks & Rules occupies a middle ground. It is deterministic except a) when multiple rules match the current state of the buffers and b) when the chunk in the buffer matches multiple chunks in a memory retrieval operation (@do get). The developer can choose whether to take advantage of this flexibility as appropriate to the use case.

Another approach is fuzzy control based upon approximate reasoning (fuzzy logic). This treats scalar values as a blend of named values, e.g. a temperature could be given as 30% cold, 60% warm and 10% hot. The fuzzy mapping is defined in terms of transfer functions for the named terms in respect to the scalar value (fuzzification).

Control is expressed using if-then rules, e.g. if it is cold, turn the heating to high; if it is warm, turn the heating to low; if it is hot, turn the heating to off. The outputs of the rules are blended based upon the input blend, and the control output determined by the reverse mapping (defuzzification). Boolean operations in rule conditions are mapped to Zadeh operators over the blend values, e.g. 30% for cold in the above example. Logical AND corresponds to the minimum such value, OR to the maximum value, and NOT to 100 minus the value when working with percentages.

FIPA (the Foundation for Intelligent Physical Agents, part of IEEE) [12] defines a framework for agent-to-agent interaction using communication acts based upon speech act theory, and enabling agents to express intentions, make requests, and share beliefs, subject to a shared ontology. FIPA further defines a variety of protocols that specify the rules and patterns for sequences of messages, ensuring orderly and predictable conversations between agents, as a basis for complex multi-step interactions like negotiations and resource allocation.

IEC 61131-3 defines a suite of graphical and textural programming languages for programmable logic controllers (PLCs) [13]. Programs can be executed once, repeatedly on a timer or on an event. The initial version of the standard was released in 1993. The 3rd edition released in 2012 features object-oriented extensions.

CAYENNE [14] is a rule-based control logic generation solution for industrial automation that introduces a rule base with domain-specific, reusable rules to automate simple, re-occurring design and implementation tasks for control logic. A rule engine applies pre-specified rules on the requirement documents, e.g. process control piping and instrumentation diagrams, to automatically generate parts of the IEC 61131-3 code.

Node-RED [15] is a NodeJS-based framework for developing event-driven distributed systems using a web-based tool for editing flow diagrams for connected nodes, and JavaScript for the code implementing the nodes. Synchronisation support includes the means for a node to wait for events from all of its input connections before proceeding.

By way of contrast, Chunks & Rules uses a simple text-based representation for facts and rules rather than structured diagrams. Actions are asynchronous, proceeding concurrently with reasoning. This mirrors the brain where conscious decisions delegate actions to the cortico-cerebellar circuit, where the cerebellum coordinates in real-time the activation of large numbers of muscles based upon sensory models in the cortex.

Learning is initially heavily dependent on cognition, but with repetition, is compiled into fast and subjectively effortless skills, e.g. riding a bicycle or playing a musical instrument. An exciting challenge for future work is to understand how we can replicate this for multi-agent neurosymbolic systems.

Acknowledgments. The author is grateful for support from the Nephele and SmartEdge projects by the European Union's Horizon Europe research and innovation programme under grant agreements No 101070487 and No 101092908, respectively.

References

1. ERCIM. https://www.ercim.eu
2. Smart Warehouse demo. https://www.w3.org/Data/demos/chunks/warehouse/
3. Nephele. https://nephele-project.eu
4. W3C Web of Things. https://www.w3.org/WoT/
5. Demo Source. https://github.com/w3c/cogai/tree/master/demos/Swarms/visualise
6. rosnodejs. https://github.com/RethinkRobotics-opensource/rosnodejs
7. ACT-R home page. http://act.psy.cmu.edu
8. Chunks & Rules spec. https://w3c.github.io/cogai/chunks-and-rules.html
9. Robot control demo. https://www.w3.org/Data/demos/chunks/robot/
10. Persimmon Group SITREP template. https://thepersimmongroup.com/situation-report-sitrep-template/
11. Langchain home page. https://www.langchain.com
12. FIPA home page. http://fipa.org
13. IEC 61131-3. https://webstore.iec.ch/en/publication/68533
14. CAYENNE. https://www.koziolek.de/docs/Koziolek2020-ICSE-SEIP-preprint.pdf
15. Node-RED. https://nodered.org

Open Access This chapter is licensed under the terms of the Creative Commons Attribution 4.0 International License (http://creativecommons.org/licenses/by/4.0/), which permits use, sharing, adaptation, distribution and reproduction in any medium or format, as long as you give appropriate credit to the original author(s) and the source, provide a link to the Creative Commons license and indicate if changes were made.

The images or other third party material in this chapter are included in the chapter's Creative Commons license, unless indicated otherwise in a credit line to the material. If material is not included in the chapter's Creative Commons license and your intended use is not permitted by statutory regulation or exceeds the permitted use, you will need to obtain permission directly from the copyright holder.

EOSC and Data Spaces for Cross-Domain Data Sharing in Europe: Insights from the TITAN-EOSC Project

Natalia Borgoñós García(✉), María Hernández Padilla, Jose Vivo Pérez, and Antonio Fernando Skarmeta Gómez

University of Murcia, Murcia, Spain
natalia.borgonosg@um.es

Abstract. Open Science is increasingly central to European research, enabling cross-sectoral access to and reuse of scientific data. To support secure, FAIR-aligned data sharing, the European Commission has launched flagship initiatives such as the European Open Science Cloud (EOSC) and the Data Spaces framework. This paper aims to discuss EOSC and Data Spaces, two of the main European initiatives designed to achieve this goal. This study will be carried out by analyzing the different strategies to achieve such a secure and cross-domain environment, as well as how these strategies can be used to contribute to the development of open science in Europe. Within this scope, the EU-funded TITAN-EOSC project will be discussed, describing its technical architecture, main objectives, and relating the key points to EOSC and Data Spaces. The main synergies between the project presented and the initiatives discussed will also be considered, highlighting how they concur with each other.

Keywords: Data Spaces · EOSC · Open Science · TITAN · FAIR Data · Data Sovereignty · Interoperability · Confidential Computing

1 Introduction

In recent years, the landscape of research and innovation in Europe is being redesigned and Open Science [1] is taking center stage. This is a global movement that aims to promote scientific research and make its results freely accessible to everyone. In other words, this movement pretends to encourage collaboration and accessibility among the scientific communities and disciplines. This transformation process contemplates the way research data is managed and shared. Taking into account that data has become a strategic asset for society over the past years, robust data management practices are crucial.

As a consequence, the European Commission [2] has been launching initiatives and strategies to enable secure data sharing across different domains and also in accordance with FAIR (Findable, Accessible, Interoperable, Reusable)

principles [3]. Among these initiatives, the European Open Science Cloud (EOSC) [4,5] is one that aims to develop a federated environment where institutions and researchers can collaborate and exchange data, as providers or consumers, in alignment with Open Science principles. [6]

In parallel, Data Spaces [6] have emerged as regulatory and technological frameworks structuring collaboration and data sharing while ensuring compliance with standards and regulations (e.g., GDPR). The emergence of Data Spaces represents a shift on how data is governed and shared.

As these initiatives are evolving, there are some projects that are funded by the European Commission that have the objective of enriching and continuing to develop them. The project TITAN-EOSC [7] (Trusted envIronments for confidenTiAl computiNg and secure data sharing) is one of them, which proposes to develop secure and trustworthy confidential data processing and sharing and demostrate them in the EOSC context. The sharing of sensitive data will follow FAIR data and open science principles, making use of the Data Spaces ecosystems. It addresses this challenge through an architectural design that combines:

- **Trusted Execution Environments (TEEs)** for confidential data processing.
- **Distributed Ledger Technologies (DLT)** for immutable access auditing.
- **Data Spaces–compliant connectors** to enable interoperable FAIR data flows.

This paper explores the convergence of Open Science, Data Spaces, and the TITAN-EOSC project. Specifically, the way that TITAN's technical architecture, based on Data Spaces infrastructure concepts, aligns with the EOSC vision. The conceptual and technical synergies will be analyzed too, in order to reflect the potential of combining the available data sharing mechanisms.

The document is organized as follows:

Section 2: Related Work. This section provides an overview of Open Science and Data Management in Europe. This will set a context for describing the conceptual framework of European Open Science Cloud (EOSC) and Data Spaces ecosystems.

Section 3: TITAN-EOSC Project. Architectural Framework. Introduces the TITAN-EOSC initiative— its origins, objectives, and technical design—and explains how it maps onto the Data Spaces concepts outlined in Sect. 2.

Section 4: TITAN-EOSC Project: Synergies between Data Spaces and EOSC. Explores the intersections between TITAN, Data Spaces, and EOSC, highlighting how the project integrates and extends these environments.

Section 5: Conclusions and Future Work. Summarizes our findings, reflects on the combined impact of these initiatives on Data Spaces and Open Science in Europe, and outlines avenues for further research.

2 Related Work

This section reviews the European Commission's main initiatives for secure, FAIR-compliant data sharing: the European Open Science Cloud (EOSC) and the emerging Data Spaces ecosystems. We present their objectives, architectures, governance models, and supporting projects.

2.1 European Open Science Cloud (EOSC)

The European Open Science Cloud (EOSC) is an initiative proposed by the European Commission and designed to provide an open environment to store and share data, encouraging scientific community to reuse research data. This is part of an European strategy to implement Open Science and develop a web of FAIR data and services for science in Europe that could be accessible to all interested researchers.

EOSC is a federation of data infrastructures that are coordinated through common standards and governance frameworks, operated by a distributed network of service providers. Its architecture is built to support interoperability, scalability, and sustainability, facilitating the integration of domain-specific repositories, cloud services and data management frameworks across disciplines and national borders.

EOSC is not a single platform but a federated infrastructure composed of three main layers [14]:

- **EOSC Core:** Essential services enabling federation—e.g., Authentication and Authorization Infrastructure (AAI) [15], Persistent Identifiers (PIDs) [16] and Metadata Catalogues.
- **EOSC Exchange** [13]: A dynamic marketplace offering onboarding workflows, data transfer services, and analysis tools.
- **EOSC Interoperability Framework** [17]: Guidelines and technical, semantic, and legal standards that ensure consistent integration across providers.

Regarding the core objectives of EOSC, it aims to address several challenges within the European research landscape. Among the key objectives of EOSC we can find:

- Enabling cross-disciplinary and cross-domain research collaborations, using a common data environment.
- Supporting the implementation of FAIR data principles.
- Establishing an interoperable ecosystem of services aligned with European standards and regulations.

EOSC has been continuously maturing from a policy vision, resulting in a structured implementation process thanks to the collaboration among european institutions, research organizations and infrastructure providers. A key point during this period was the creation of the EOSC Association [8], the legal entity that governs and coordinates the EOSC's strategic agenda and representing its stakeholder community. This association was created in 2020 and works alongside the European Commission and the European Member States and associated countries in a tripartite governance model [9] to resource and support the implementation of the EOSC environment in Europe, advance an Open Science system and aligning national and EU policies to improve the production of FAIR research output.

The implementation of EOSC [12] is based on a long-term process of alignment and coordination pursued by the Commission since 2015. This has involved a diverse range of stakeholders in the European research landscape. This process refers to the transition from the initial policy vision into a federated ecosystem that provides European researchers with access to data and services, being guided by strategic roadmaps and supported by European funding.

Implementation is driven by numerous EU-funded projects, under Horizon 2020 [10] and Horizon Europe [11] research and innovation programs, which play a crucial role in translating the EOSC policy vision into a functional and sustainable ecosystem by developing its core components. These projects contribute to developing the EOSC Core infrastructure that includes the essential capabilities (identity management, metadata catalogs, ...) and enhancing the EOSC Exchange [13] with domain specific assets.

Some outstanding projects supporting EOSC are:

- **EOSC Future.** [18] Builds and consolidates EOSC Core, integrating foundational services and fostering standards adoption. Creates the EOSC Portal as a single access point in order to enable researchers to access FAIR-compliant data across Europe.
- **FAIRsFAIR.** [19] Implements FAIR principles across research workflows via tools, training, and best practices.
- **EOSC ENTRUST.** [20] The objective of this project is to enhance the different federated identity and trust mechanisms within the EOSC Context. Designs a network of Trusted Research Environments (TREs) [22] for federated access to sensitive data, aligning TREs with EOSC and Data Spaces standards.

2.2 Data Spaces Ecosystems

Driven by Europe's data strategy, the European Data Spaces initiative [23] is building sovereign, trusted, and interoperable frameworks for cross-sector data sharing. These collaborative ecosystems let organizations securely share, manage, and utilize data while retaining full ownership and control. In contrast to conventional sharing methods, Data Spaces embed trust through clear governance models and technical standards, enabling a federated exchange in which each participant maintains data sovereignty, transparency, and accountability.

Core benefits of Data Spaces:

- **Decentralized data exchange:** Each participant keeps data in its infrastructure.
- **Data sovereignity:** Data owners decide on how and when data is used to control it.
- **Data Value Creation:** Contributing to new solutions to develop responsible data usage.

To align efforts across Europe's data economy, the Data Spaces Business Alliance (DSBA) [26] brings together four leading bodies: Big Data Value Association (BDVA) [28], FIWARE Foundation [29], Gaia-X European Association for Data and Cloud [24,30] and International Data Spaces Association (IDSA) [31].

Under DSBA's umbrella, the Technical Convergence Document [27] defines a unified Blueprint Architecture [32] composed of modular Technical Building Blocks [33]. These building blocks—covering components such as catalogs, semantic models, governance APIs, and security services—ensure that independent Data Spaces can interoperate, scale, and maintain trust.

The Data Spaces Support Centre (DSSC) [34], funded by the Digital Europe Programme [35], translates the Blueprint into practice. It offers detailed blueprints and reference implementations, training materials and best-practice guides and alignment with EU policy goals and regulatory requirements. The architecture of a data space is built around several core concepts that ensure the system is decentralized, secure, and interoperable. A data space is built upon the previously mentioned Technical Building Blocks, that help to create a modular, scalable, and interoperable system, enabling secure data exchange and collaboration among different entities [41]. These building blocks are based on different tasks that are needed and integrated to create a functional data space.

A typical Data Space architecture, see Fig. 1 splits into a Control Plane—handling identity, policy, and governance—and a Data Plane, which manages FAIR-compliant data flows. At the heart of this design sits the Connector component [36], responsible for mediating secure data transfers between participants while enforcing access-control and usage-control policies according to owner stipulations. Furthermore, applying sovereignty rules (e.g., geographic, temporal restrictions) and interfacing with trust frameworks and identity infrastructures are other functions that this element also performs.

Connectors [25] are fundamental elements in a Data Space, as they enable secure access, exchange, and control of data in compliance with the policies established by the data owners. They constitute a standardized point of interaction among the participants in the space.

Connectors encapsulate the ecosystem's roles [38,39] (see Fig. 2) by providing a standardized interaction point for:

- **Data Providers:** Publish metadata and datasets to catalogs [40].
- **Data Owners:** Authorize or restrict consumer access.
- **Data Consumers:** Discover and consume datasets under agreed terms.
- **Data Intermediaries:** Offer value-added services (e.g., quality assurance, transformation).

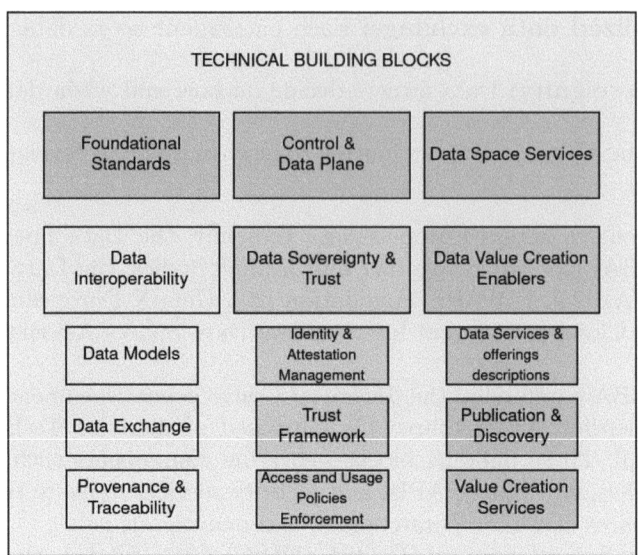

Fig. 1. Technical Building Blocks.

- **Technology providers:** Supply and maintain underlying infrastructure and tools.

A Data Space could count on different components depending on the approach adopted. Despite of this, there are some of them that are strictly necessary for a Data Space to perform its basic activity as required. One of these components and one of the main elements in the context of data spaces is the previously mentioned Connector.

FIWARE has developed its own implementation of a connector [37] to seamlessly integrate within its ecosystem. This implementation [23] adheres to data sovereignty principles and ensures that users can share information securely and in compliance with defined access policies. It is a Base Connector that enables data exchange between participants in a data space, which can be extended or integrated with additional tools to meet stricter requirements related to security or compliance.

3 Project TITAN-EOSC. Architectural Framework

The project TITAN-EOSC, funded by the European Commission, aims mainly to enrich the EOSC Interoperability Framework (EOSC-IF) by developing a software platform solution for confidential collaboration and privacy-preserving data processing.

3.1 Objectives

This is to be achieved by meeting the following steps and objectives:

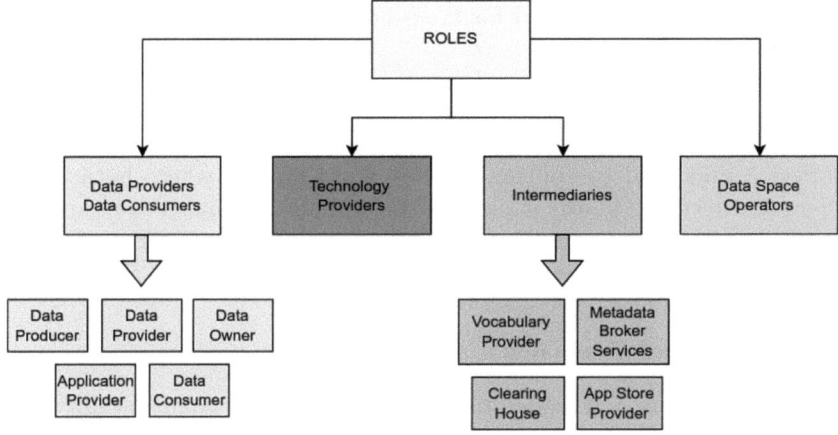

Fig. 2. Roles involved in a Data Space.

- Collect legal, technical and architectural requirements. Then, define a platform architecture for secure sharing of sensitive data and publishing anonymised datasets in the EOSC-IF.
- Develop secure data sharing, auditing mechanisms for sensitive data and including secure data zones, data access control and end-to-end data protection.
- Develop an end-to-end secure data processing framework for collaborative and privacy-preserving Machine Learning (ML) [43] using Trusted Execution Environments [42].
- Implement confidential mechanisms, algorithms and tools with cloud infrastructure platforms and the EOSC IF, and validate solutions in sensitive data-driven use cases, especifically, government and healthcare.
- Disseminate and promote the solutions for data governance and stewardship through collaboration with EOSC Partnership initiatives, standardisation and integrating with the EOSC infrastructure.

TITAN-EOSC enhances the European Open Science Cloud by enabling secure, privacy-preserving access to sensitive datasets held by governmental, public, and private entities—with a focus on government administration and healthcare use cases. The platform ensures compliance with GDPR [47], the European Digital Identity Framework [48], and other applicable regulations.

To mitigate security risks inherent in data sharing, TITAN-EOSC leverages:

- **Trusted Execution Environments (TEEs).** Hardware-backed enclaves for in-use data confidentiality. [44]
- **Privacy-Enhancing Technologies (PETs).** Reliable anonymization and privacy tools for dataset sanitization [54].
- **Distributed Ledger Technologies (DLT).** Decentralized, tamper-evident access control and audit logging [53].

These components are designed for cross-border demonstrations and full compatibility with EOSC-IF.

TITAN-EOSC's success depends on harmonizing Europe's legal framework for data protection with cutting-edge technologies from Confidential Computing (CC) [45], PETs, Machine Learning (ML) [43], and DLT domains. TEE-supported confidential computing enables novel models where both code and data remain encrypted in use, while DLT-backed access management and remote attestation provide cryptographically verifiable proofs of execution integrity and transparency.

3.2 Architectural Overview

TITAN's architecture is designed to seamlessly integrate confidentiality, policy-driven control, and FAIR data management into the EOSC-IF and Data Spaces paradigms. As depicted in Fig. 3, the architecture comprises three intertwined layers—Infrastructure, Control Plane, and Data Plane—each hosting a set of components that collaborate to ensure end-to-end security, interoperability, and usability.

Infrastructure. This foundational layer provides the shared "plumbing" upon which higher-level services operate:

- **Trust Authority.** The recognized governance body or consortium that issues, revokes, and oversees trust anchors—identities, certificates, and governance policies. Bootstraps the entire ecosystem by provisioning initial credentials for all other components.
- **Authentication and Authorization (AA) Infrastructure.** Self-Sovereign Identity (SSI) frameworks, Decentralized Identifiers (DIDs), and the Distributed Ledger–backed Verifiable Data Registry (VDR)- acting as an immutable ledger recording DIDs, credential schemas, revocation lists, and governance policies.
- **Catalog and Agreement Manager.** The catalog houses dataset metadata (e.g., titles, descriptions, PIDs), FAIR metrics, and pointers to physical storage. At the same time, the Agreement Manager records consent frameworks and usage agreements at dataset level, linking them to catalog entries and encoding them as machine-readable policies.
- **Data Anonymisation Assessment Service.** Evaluates re-identification risk and produces anonymisation reports. These reports, along with sanitized data descriptions, are cross-referenced in the Catalog to prepare datasets for secure sharing.
- **TITAN Dashboard.** A unified portal that aggregates discovery, policy configuration, anonymisation tools, and compute job launchers. It serves as the primary user interface for both Providers and Consumers.

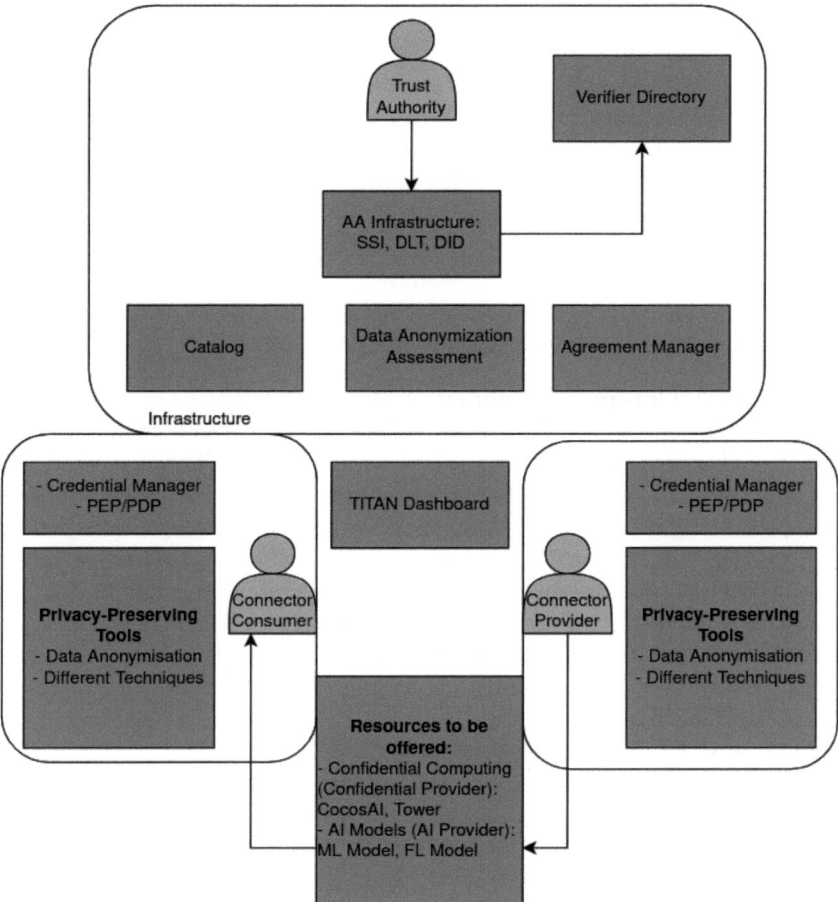

Fig. 3. TITAN-EOSC Architecture.

Control Plane. The Control Plane enforces all governance, policy, and identity workflows:

- **Verifiable Data Registry (VDR).** Although physically part of the Infrastructure, its logical role in policy retrieval and credential verification places it at the heart of Control Plane operations. Every policy lookup or credential proof originates here.
- **Data Space Connector**:
  - **Policy Enforcement Point (PEP).** Intercepts all incoming requests for data or compute services.
  - **Policy Decision Point (PDP).** Evaluates requests against policies retrieved from the VDR and Agreement Manager, taking into account user attributes, dataset usage agreements, and contextual constraints (e.g., time, geography).

- **Credential Manager**. Interfaces with user SSI Wallets to validate presented credentials, extract minimal required claims, and submit proofs for PDP evaluation—ensuring attribute disclosure follows the principle of data minimization.
- **Data Transfer Coordination**. Negotiates secure data flows using standardized protocols (e.g., IDS Messaging) and ensures data endpoints comply with usage policies.

Data Plane. The Data Plane hosts the actual computation and storage resources that deliver privacy-preserving analytics:

- **Confidential Computing Services** [46]: Ensuring security during operational execution, where Hardware infrastructure, which includes Trusted Execution Environments (TEE), plays a critical role. These trusted execution environments create a safe and isolated space for processing confidential data, protecting it from unauthorized access and minimizing the risk of leaks.
  - **COCOS AI**. A containerized TEE runtime where AI/ML workloads execute with in-memory encryption. It supports model training and inference without exposing raw data or model parameters to the host OS.
  - **Tower**. A resource orchestrator that provisions TEE-enabled VMs or containers on demand, dynamically scaling computing capacity in response to workload requirements.
- **AI/ML and FL Model Repositories**:
  - **ML Models**. Pre-trained or user-submitted models for specific domains (e.g., agrifood, sleep analysis) that run within confidential environments.
  - **Federated Learning (FL) Module**. Coordinates multi-party model training across distributed TEEs, aggregating secure model updates via the VDR to preserve individual data privacy.
- **Privacy-Preserving Tools**: A suite of plug-and-play modules—data anonymisation algorithms, encryption libraries—that Providers and Consumers can invoke directly through their Connectors to sanitize or transform datasets prior to compute tasks.

3.3 End-to-End Workflow

This subsection outlines the complete sequence of interactions within the TITAN platform, illustrating how data moves securely from provider to consumer under strict confidentiality, policy enforcement, and FAIR publication guarantees.

1. **Dataset Onboarding**: A Data Provider registers a dataset in the Catalog; usage terms are recorded by the Agreement Manager.
2. **Access Request**: A Data Consumer initiates a request via their Connector. The Privacy and Security Manager retrieves their verifiable credentials from the VDR and the PDP evaluates policies.

3. **Secure Processing**: Anonymization jobs run to produce sanitized outputs. Confidential ML/FL tasks execute inside TEEs (COCOS AI/Tower), with remote attestation ensuring code integrity.
4. **Auditing and Logging**: All authentication events, policy decisions, and TEE attestations are immutably recorded on the DLT and discoverable via the Verifier Directory.
5. **Result Publication**: Anonymized datasets or model artifacts are published back to the Catalog with FAIR metadata and PIDs, ready for downstream reuse in EOSC-IF.

4 Project TITAN-EOSC: Synergies Between Data Spaces and EOSC

TITAN sits at the intersection of two complementary European data-sharing paradigms: Data Spaces—often tailored to specific domains—and the federated, FAIR-driven European Open Science Cloud (EOSC). By combining the governance and interoperability capabilities of Data Spaces with EOSC's open-science ethos, TITAN delivers a unified platform that enforces confidentiality, trust, and sovereign control over research data.

4.1 Converging Visions

- **Federation and Interoperability**: EOSC and Data Spaces both champion a distributed, standards-based infrastructure in which institutions retain local control of data while contributing to a larger, searchable ecosystem. TITAN's connector-centric design leverages common protocols and metadata models to seamlessly integrate with both EOSC catalogues and Data Spaces directories, enabling cross-domain discovery and data exchange.
- **FAIR and Sovereign Data Sharing**: EOSC promotes FAIR principles; Data Spaces embed data-sovereignty guarantees. TITAN enforces FAIRness at every step—dataset registration, metadata publication, access control, and result sharing—while ensuring that data owners define and enforce their own usage policies through verifiable credentials and policy-decision components.

4.2 Core Synergy Pillars

TITAN's architecture crystallizes three essential synergies between Data Spaces and EOSC:

- **Interoperability.** Through standardized formats, protocols, and connector-driven architecture, TITAN's Data Spaces and EOSC enable seamless integration of data from diverse sources and domains. This interoperability empowers researchers to combine datasets across disciplines and borders, driving interdisciplinary collaboration and pooling resources to tackle global challenges.

By unlocking access to FAIR-compliant, large-scale repositories, the platform fosters more robust scientific methodologies, accelerates novel discoveries, and catalyzes the development of advanced tools and technologies for comprehensive research outcomes.
- **Data sovereignty.** By enabling owners to manage and enforce access and usage policies in a decentralized, granular fashion, data sovereignty ensures that researchers and institutions retain full control over their assets. As a cornerstone of both TITAN and EOSC, this principle lets data providers specify exactly who may use their data and under what conditions—building trust among participants and encouraging broader, more open collaboration.
- **Trust frameworks.** By employing verifiable credentials and decentralized identity schemes—coordinated through the connector—TITAN and EOSC guarantee secure, auditable, and trustworthy data transactions. In light of Europe's strict data-protection regulations, these frameworks ensure that all sharing and usage comply with legal requirements, helping researchers avoid compliance risks and safeguard sensitive information.

4.3 Supporting EOSC Growth

As EOSC grows, it needs a scalable and sustainable infrastructure to manage the increasing amounts of data. Data spaces provide a decentralized but interconnected environment, where data can be managed efficiently across various platforms and institutions. This scalability is crucial for the long-term sustainability of the EOSC. Data spaces are crucial for EOSC and the TITAN-EOSC project, providing the infrastructure needed to support secure, interoperable, and compliant data sharing across Europe.

5 Conclusions and Future Work

By integrating Data Spaces' governance and sovereignty features with EOSC's federated, FAIR-driven infrastructure, the TITAN-EOSC project delivers a unified platform for confidential, privacy-preserving data processing and collaborative research. Our connector-centric framework—anchored in Trusted Execution Environments (TEEs), self-sovereign identity, and decentralized blockchain-backed access control—ensures end-to-end security, transparent auditability, and fine-grained policy enforcement. Demonstrated in public administration and healthcare use cases, TITA-EOSCC enriches the EOSC Interoperability Framework with scalable mechanisms for managing sensitive datasets across borders and domains.

Looking ahead, we identify four key directions for future work:

1. **Semantic Interoperability**: Continue to align with emerging European efforts on shared vocabularies, ontologies, and metadata standards. By reusing community-recognized schemes and contributing to common semantic frameworks, TITAN-EOSC can further reduce barriers to cross-domain data understanding and integration.

2. **Expanded Pilots and FAIR Metrics**: Scale deployments into additional sectors—such as energy, environment, and smart cities—to validate cross-domain interoperability in real-world settings. Integrate automated FAIR-metric assessments and provenance tracking to provide on-the-fly compliance and reuse readiness indicators.
3. **Standards and Ecosystem Alignment**: Deepen collaboration with standardization bodies (e.g., DSSC, EOSC Association) to ensure that TITAN's connectors, policies, and audit mechanisms map directly to evolving EU Data Spaces and EOSC specifications. Participation in working groups will drive broad adoption and long-term sustainability.
4. **Ethical, Privacy-Centered Innovation**: Explore advanced Privacy-Enhancing Technologies (homomorphic encryption, secure multi-party computation) and adaptive consent models to address evolving regulatory landscapes and ethical considerations around data reuse.

By pursuing these avenues, TITAN-EOSC will continue to strengthen Europe's open-science ecosystem—promoting transparency, reproducibility, and responsible data reuse—while safeguarding individual and institutional sovereignty over sensitive research assets.

Acknowledgements. This paper has been funded by the European Union under the GA No 101129822, for the TITAN-EOSC Project. Views and opinions expressed are however those of the author(s) only and do not necessarily reflect those of the European Union or the European Research Executive Agency (REA). Neither the European Union nor the granting authority can be held responsible for them.

References

1. Open Science. https://research-and-innovation.ec.europa.eu/strategy/strategy-research-and-innovation/our-digital-future/open-science_en. Accessed 15 May 2025
2. European Commission. https://commission.europa.eu/index_en. Accessed 15 May 2025
3. The FAIR Guiding Principles for scientific data management and stewardship. https://www.nature.com/articles/sdata201618. Accessed 15 May 2025
4. European Open Science Cloud (EOSC). https://research-and-innovation.ec.europa.eu/strategy/strategy-research-and-innovation/our-digital-future/open-science/european-open-science-cloud-eosc_en. Accessed 15 May 2025
5. EOSC. https://eosc.eu/. Accessed 15 May 2025
6. What is a Data Space?. https://language-data-space.ec.europa.eu/help/faq/what-data-space_en. Accessed 15 May 2025
7. Titan EOSC Project. https://titan-eosc.eu. Accessed 15 May 2025
8. EOSC Association. https://eosc.eu/eosc-association/. Accessed 15 May 2025
9. Tripartite Collaboration. https://eosc.eu/tripartite-collaboration/. Accessed 15 May 2025
10. Horizon 2020. https://research-and-innovation.ec.europa.eu/funding/funding-opportunities/funding-programmes-and-open-calls/horizon-2020_en. Accessed 15 May 2025

11. Horizon Europe. https://research-and-innovation.ec.europa.eu/funding/funding-opportunities/funding-programmes-and-open-calls/horizon-europe_en. Accessed 15 May 2025
12. Budroni, P., Claude-Burgelman, J., Schouppe, M.: Architectures of knowledge: the european open science cloud. ABI Technik **39**(2), 130–141 (2019). https://doi.org/10.1515/abitech-2019-2006, last accessed 2025/08/20
13. EOSC Exchange. https://eoscfuture.eu/ker/eosc-exchange/. Accessed 16 May 2025
14. Architecture of EOSC. https://handbook.eosc-synergy.eu/eosc-architecture/. Accessed 16 May 2025
15. EOSC Authentication and Authorization Infrastructure (AAI). https://op.europa.eu/en/publication-detail/-/publication/d1bc3702-61e5-11eb-aeb5-01aa75ed71a1. Accessed 16 May 2025
16. A Persistent Identifier (PID) policy for the European Open Science Cloud (EOSC). https://op.europa.eu/es/publication-detail/-/publication/35c5ca10-1417-11eb-b57e-01aa75ed71a1/language-en. Accessed 16 May 2025
17. EOSC Interoperability Framework. https://op.europa.eu/en/publication-detail/-/publication/d787ea54-6a87-11eb-aeb5-01aa75ed71a1/language-en. Accessed 16 May 2025
18. Project EOSC Future. https://eoscfuture.eu. Accessed 16 May 2025
19. Project FAIRsFAIR. https://www.fairsfair.eu/. Accessed 19 May 2025
20. Project EOSC ENTRUST. https://www.eosc-entrust.eu. Accessed 19 May 2025
21. Project FAIR-IMPACT. https://www.fair-impact.eu/. Accessed 19 May 2025
22. What are Trusted Research Environments?. https://www.researchdata.scot/engage-and-learn/data-explainers/what-are-trusted-research-environments/. Accessed 19 May 2025
23. Otto, B., ten Hompel, M., Wrobel, S.: Designing Data Spaces. Springer Cham (2022). https://doi.org/10.1007/978-3-030-93975-5. Accessed 20 Aug 2025
24. Otto, B.: A Federated Infrastructure for European Data Spaces. Association for Computing Machinery (2022). https://doi.org/10.1145/3512341. Accessed 20 Aug 2025
25. Gieß, A., et al.: What does it take to connect? Unveiling characteristics of data space connectors. In: Proceedings of the 57th Hawaii International Conference on System Sciences (2024). https://doi.org/10.24251/HICSS.2024.511, last accessed 2025/08/20
26. The Data Spaces Business Alliance. https://data-spaces-business-alliance.eu/. Accessed 19 May 2025
27. The Data Spaces Business Alliance: Technical Convergence. Discussion Document. https://data-spaces-business-alliance.eu/wp-content/uploads/dlm_uploads/Data-Spaces-Business-Alliance-Technical-Convergence-V2.pdf. Accessed 20 May 2025
28. Big Data Value Association. https://bdva.eu/. Accessed 20 May 2025
29. FIWARE, the Open Source Platform for Our Smart Digital Future. https://www.fiware.org/. Accessed 20 May 2025
30. GAIA-X. https://gaia-x.eu/. Accessed 20 May 2025
31. International Data Spaces Association. https://internationaldataspaces.org/. Accessed 20 May 2025
32. Data Spaces Blueprint v1.5. https://dssc.eu/space/bv15e/766061169/Data+Spaces+Blueprint+v1.5+-+Home. Accessed 20 May 2025
33. Building-Blocks-Catalog. https://docs.internationaldataspaces.org/ids-knowledgebase/open-dei-building-blocks-catalog. Accessed 20 May 2025

34. Data Spaces Support Centre. https://dssc.eu/. Accessed 20 May 2025
35. The Digital Europe Programme. https://digital-strategy.ec.europa.eu/en/activities/digital-programme. Accessed 20 May 2025
36. IDS Connector. https://docs.internationaldataspaces.org/ids-knowledgebase/ids-ram-4/layers-of-the-reference-architecture-model/3-layers-of-the-reference-architecture-model/3_5_0_system_layer/3_5_2_ids_connector. Accessed 20 May 2025
37. FIWARE Data Space Connector. https://github.com/FIWARE/data-space-connector. Accessed 20 May 2025
38. Roles in the International Data Spaces. https://docs.internationaldataspaces.org/ids-knowledgebase/ids-ram-4/layers-of-the-reference-architecture-model/3-layers-of-the-reference-architecture-model/3-1-business-layer/3_1_1_roles_in_the_ids. Accessed 20 May 2025
39. What are the main elements of a data space?. https://datos.gob.es/en/blog/what-are-main-elements-data-space. Accessed 20 May 2025 https://docs.internationaldataspaces.org/ids-knowledgebase/dataspace-protocol/overview/terminology
40. Catalog Protocol. https://docs.internationaldataspaces.org/ids-knowledgebase/dataspace-protocol/catalog/catalog.protocol. Accessed 21 May 2025
41. Terminology. https://docs.internationaldataspaces.org/ids-knowledgebase/dataspace-protocol/overview/terminology. Accessed 21 May 2025
42. Trusted Execution Environment. https://dualitytech.com/glossary/trusted-execution-environment/. Accessed 22 May 2025
43. Xu, R., Baracaldo, N., Joshi, J.: Privacy-Preserving Machine Learning: Methods, Challenges and Directions. arXiv preprint arXiv:2108.01426 (2021). Last revised: 22 Sep 2021 (v2), last accessed 2025/05/22. Available at: https://arxiv.org/abs/2108.01426
44. Sabt, M., Achemlal, M., Bouabdallah, A.: Trusted Execution Environment: What It is, and What It is Not, IEEE Trustcom/BigDataSE/ISPA (2015). https://doi.org/10.1109/Trustcom.2015.357. Accessed 22 May 2025
45. What is confidential computing?. https://www.ibm.com/think/topics/confidential-computing. Accessed 22 May 2025
46. Feng, D., et al.: Survey of research on confidential computing (2015). https://doi.org/10.1049/cmu2.12759. Accessed 20 Aug 2025
47. General Data Protection Regulation (GDPR). https://gdpr.eu. Accessed 22 May 2025
48. European Digital Identity (EUDI) Regulation. https://digital-strategy.ec.europa.eu/en/policies/eudi-regulation. Accessed 22 May 2025
49. What is a Policy Enforcement Point (PEP)?. https://www.nextlabs.com/blogs/what-is-a-policy-enforcement-point-pep/. Accessed 23 May 2025
50. Self-Sovereign Identity. https://seon.io/resources/dictionary/self-sovereign-identity/. Accessed 23 May 2025
51. walt.id: Digital identity and wallet infrastructure. https://walt.id. Accessed 23 May 2025
52. What is a field programmable gate array (FPGA)?. https://www.ibm.com/think/topics/field-programmable-gate-arrays. Accessed 23 May 2025
53. What is distributed ledger technology (DLT)?. https://www.techtarget.com/searchcio/definition/distributed-ledger. Accessed 23 May 2025
54. Privacy-Enhancing Technologies. https://royalsociety.org/news-resources/projects/privacy-enhancing-technologies/. Accessed 23 May 2025

55. EOSC in the Data Spaces frame. https://eosc.eu/wp-content/uploads/2023/06/03_20230626_EOSC_in_the_Data_Spaces_Frame_KL.pdf. Accessed 26 May 2025
56. DATA SPACES SUPPORT CENTRE. https://eosc.eu/wp-content/uploads/2023/06/02_DSSC_presentation_Savvas.pdf. Accessed 26 May 2025

Open Access This chapter is licensed under the terms of the Creative Commons Attribution 4.0 International License (http://creativecommons.org/licenses/by/4.0/), which permits use, sharing, adaptation, distribution and reproduction in any medium or format, as long as you give appropriate credit to the original author(s) and the source, provide a link to the Creative Commons license and indicate if changes were made.

The images or other third party material in this chapter are included in the chapter's Creative Commons license, unless indicated otherwise in a credit line to the material. If material is not included in the chapter's Creative Commons license and your intended use is not permitted by statutory regulation or exceeds the permitted use, you will need to obtain permission directly from the copyright holder.

Multi-Agent Stateless Orchestration for Distributed Data Pipelines Implementation

Nicolò Bertozzi[✉][iD], Anna Geraci[iD], Marco Sacchet[iD], Enrico Ferrera[iD], and Claudio Pastrone[iD]

Fondazione Links, Via Pier Carlo Boggio 61, 10138 Turin, Italy
{nicolo.bertozzi,anna.geraci,marco.sacchet,enrico.ferrera, claudio.pastrone}@linksfoundation.com

Abstract. This paper introduces a stateless orchestration architecture and mechanism for managing distributed data pipelines in multi-agent systems. Workflows are decomposed into independent subworkflows by a splitting module. A caching mechanism preserves execution context without relying on local state, supporting true statelessness. Communication is event-driven via Apache Kafka, and a unified data model ensures consistent interaction among components. The architecture is validated on a Kubernetes environment, showing scaling and low-latency features. This approach combines the control of orchestration with the scalability of event-driven systems, offering a robust solution for modular workflow execution.

Keywords: Stateless Orchestration · Multi-Agent Systems · Asynchronous Communication · Distributed Architecture

1 Introduction

This work extends the previous short paper [2], in which the authors presented a solution to ease the deployment of industrial digital twins while enhancing operational efficiency. The idea behind the aforementioned architecture was to facilitate the integration of IoT Sensors and AI modules with digital twins' simulations. The implementation in real-world scenarios has highlighted the need for some additional features and tweaks. In the present paper, the authors aim to illustrate the most recent advancements made in the development of this architecture and, at the same time, to delve into certain aspects that could not be adequately addressed in our initial publication.

2 Related Work

In [2], the literature on this matter was extensively analysed, and what was presented there is extended in this work. This extension is developed in the context of modern distributed systems, where the need to define and coordinate the execution of multiple agents has become increasingly common. This coordination is typically formalised as a workflow, a sequence of interdependent steps involving multiple actors, each contributing to a shared goal. The structure and communication model of such workflows can significantly affect performance, scalability, and fault tolerance. This study explores various strategies for workflow management, drawing on insights from the comparative analysis presented in [11]. That study focuses on two widely adopted coordination paradigms: event choreography and orchestration. Event choreography, as shown in [4,8] is a decentralised model in which individual components independently execute their operations and publish events to signal state changes. They perform a reactive chain of events without any central controller, providing also the possibility of integration with other technologies, as shown in [4]. Orchestration, by contrast, is a centralised model where a dedicated orchestrator governs the sequence of operations. It listens to events from distributed services, maintains a global view of the workflow, and explicitly invokes each step in the process according to a predefined logic. This approach is implemented in a large variety of different use cases, as shown in [5,6,13]. The referenced study, followed also by [1,12], evaluates both models and finds that, while event choreography achieves significantly lower execution times compared to orchestration, orchestration offers superior clarity, traceability, and control; particularly in complex workflows. These findings highlight the inherent trade-off between performance and manageability in workflow coordination. The orchestrator reasoning capability to understand and execute the tasks described within the workflows is essential.

The paper [10] introduces a modular architecture in which a central planner decomposes complex tasks into sub-tasks and distributes them to a pool of agents, each specialised in solving specific problems. The planner is designed to select among alternative strategies based on task characteristics, performance constraints, and available resources. It can generate execution paths either sequentially or in parallel, dynamically choosing the optimal agents to achieve the desired goal. Inspired by this notion of intelligent planning [14], the proposed work integrates a preprocessing component that serves a similar function by transforming high-level user intents into executable workflow steps. This preprocessing module is specifically tailored to the platform's requirements and enables efficient, scalable orchestration logic, while preserving the flexibility to incorporate more advanced planning strategies, including those powered by GenAI or large language models (LLMs), in future iterations. In this context, while the presented system adopts a lightweight, programmatic approach to workflow definition to maximise simplicity, we acknowledge the value of formal models like BPMN for long-term maintainability and interoperability [3,7]. As shown in [9], integrating standardised notations in orchestration engines can enhance monitoring and runtime adaptability. Building on this understanding, our work focuses specifically on orchestration, adopting a centralised model tailored to meet the specific requirements of our platform. Our primary objective is to enhance latency performance while preserving the inherent advantages of centralised orchestration, adding a stateless design that enables scalability. Unlike also other orchestration frameworks such as AWS Step Functions and Netflix Conductor, which retain local state, this represents a novelty in the current state of the art. The design, operation, and role of our orchestration component are detailed in later sections.

3 Architecture

A brief description of the architecture shown in Fig. 1 is provided with a short explanation of every component. The Pipeline Management macro-block, though not the focus of this work, is included for completeness, as it offers auxiliary features that enhance but are not essential for running the platform.

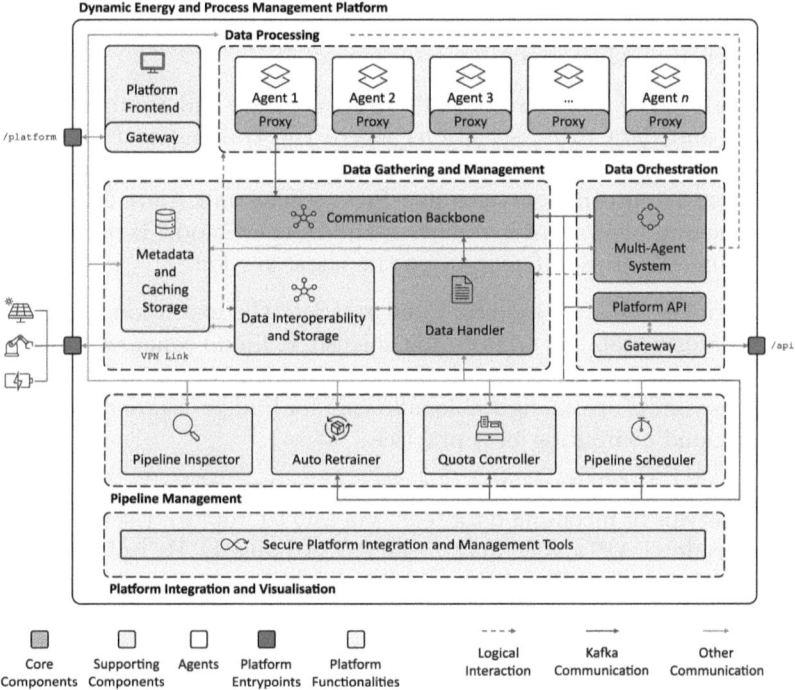

Fig. 1. The architecture of the Multi-Agent System environment.

Multi-Agent System (MAS). It acts as the orchestrator, triggering agents asynchronously in response to workflows. It interfaces with Agents and Data Handler (DH), and logs state in Metadata and Caching Storage (MCS) for resilience.

Communication Backbone (CB). It manages asynchronous event-based communication among platform components and user inputs. Built on Apache Kafka, it ensures robust message delivery.

Agent and Proxy. Agents execute user-defined tasks, while the Agent Proxies (AP) abstracts communication with CB, MCS, and Data Interoperability and Storage (DIS), allowing agents to remain focused on computation.

Data Handler (DH). It provides an interface between the DIS and core services, handling data retrieval and decoupling storage formats for flexibility.

Platform APIs (PAPIs). It organises RESTful APIs, with role-based access for users, administrators, and managers.

Platform Frontend. Provides the GUI to the user, integrating with PAPI for visualisation of simulations, workflows, and configurations.

Pipeline Inspector (PI). It is a web interface that allows platform managers to visualise and troubleshoot message flows across the system. It analyses Apache Kafka topics to identify anomalies in payload structure or agent assignment. This insight supports quick debugging and aids in updating agents or alerting developers when pipeline execution issues arise.

Auto Retrainer (AR). It ensures agents stay up to date and accurate by addressing concept and data drift, situations where the training and inference data distributions diverge. It follows MLOps principles to automate retraining based on alerts from monitoring services like Prometheus and Alert Manager. Once retrained, the outdated agent is removed, and the new model is published to the MCS, ensuring all future inferences use the latest version.

Quota Controller (QC). It monitors and restricts the execution of pipelines to maintain platform stability and prevent misuse. It limits concurrent executions, protecting against excessive use during development or potential brute-force attacks. These limits are configurable, allowing for budget-aware resource control inspired by cloud computing best practices.

Pipeline Scheduler (PS). It enables automated pipeline triggering based on either temporal intervals or incoming data (e.g., via MQTT topics). Users configure the scheduler through PAPIs using authorised POST requests. By isolating schedules to approved workflows, PS ensures secure and predictable task automation.

4 Data Model

Figure 2 presents the data models exchanged over the CB, each transmitted via a dedicated Kafka topic that links specific platform components. The *orchestrator topic* enables user-to-MAS interactions by carrying `Orchestrator Tasks`; the *agent topic* connects MAS and Agents through `Agent Tasks`; and the *data topic* links the DH with the MAS via `Data Tasks`.

All messages share core fields such as `id`, `state`, `issuer`, and `timestamp`, which are used to track task progress and reconstruct pipeline execution histories. Further details on the structure and purpose of each message type are provided in the next section.

```
Agent Task
{
    «state»: <PLANNED|PAUSED|COMPLETED|STOPPED|ERROR>,
    «phase»: <INFERENCE|TRAINING>,
    «id»: <task id>,
    «agent»: <agent id>,
    «issuer»: <user group that used the Platform APIs>,
    «timestamp»: <timestamp of generation>,
    «timeout»: <maximum time allowed for the execution>,
    «data»: <optional configuration data>,
    «data_source»: <optional data source to contact> {
        «url»: <url to contact>,
        «public_key»: <agent public key>
    },
    «task_restart»: <following split of the workflow>,
    «variables»: <variables of the last workflow split>
}
```

```
Scheduler Task
{
    «request»: <PLANNED|CANCELLED>,
    «trigger»: <TIME|DATA>,
    «state»: <ACCEPTED|REJECTED|ERROR>,
    «id»: <task id>,
    «issuer»: <user group that used the Platform APIs>,
    «timestamp»: <timestamp of generation>,
    «start_time»: <start time of the schedule>,
    «include_start»: <include start_time as trigger>,
    «repeat_pattern»: <pattern for scheduling events>,
    «topic»: <MQTT topic that triggers the event>,
    «workflow»: <workflow to execute>
}
```

```
Orchestrator Task
{
    «state»: <PLANNED|COMPLETED|STOPPED|ERROR>,
    «id»: <task id>,
    «issuer»: <user group that used the Platform APIs>,
    «timestamp»: <timestamp of generation>,
    «timeout»: <maximum time allowed for the execution>,
    «schedule_id»: <optional corresponding schedule>,
    «workflow»: <name of the workflow>,
    «data»: <optional configuration data>,
}
```

```
Data Task
{
    «state»: <REQUEST|RESPONSE>,
    «id»: <data task id>,
    «issuer»: <user group that used the Platform APIs>,
    «timestamp»: <timestamp of generation>,
    «data_source»: <agent task data souce> {...}
    «data»: <optional configuration data>,
    «task_restart»: <following split of the workflow>,
    «variables»: <variables of the last workflow split>
}
```

```
Quota
{
    «user»: <user that has used the Platform APIs>,
    «issuer»: <user group that used the Platform APIs>,
    «resource»: <PIPELINE|SCHEDULE|DEPLOYMENT|CONFIG>,
    «timestamp»: <timestamp of generation >
}
```

Fig. 2. The set of data models adopted in the platform to enable a standardised communication among the components.

Orchestrator Task. It initiates a new data pipeline execution, specifying the workflow name and optional input data. If triggered by the PS, a schedule_id is included. Users may define a timeout, defaulting to 600 s, which is slightly above the platform's maximum estimated execution time.

Agent Task. It includes any required input for data requests. Hence, it provides a pointer to the source that the DH must access. The task_restart field informs the MAS where to resume execution upon completion and supports the stateless design discussed later in Sect. 5.2. This logic is shared with the Data Task.

Data Task. It carries a reference to the data_source to be queried, optionally duplicated in the input field. It also includes the task_restart field with the same function as in the Agent Task.

Scheduler Task. It handles both time- and data-based triggers, distinguished by the corresponding key (trigger). Unlike other tasks, its state field indicates whether the requested scheduling action was accepted. The workflow field stores the file name to be executed, while start_time, include_start, and repeat_pattern apply to *time* triggers, and topic applies to *data* triggers.

Quota. It tracks user operations via a lightweight monitoring system that logs `user` identity, `issuer`, operation type (`resource`), and `timestamp`. It helps detecting excessive or scripted activity, acting as a safeguard against misuse or malicious access and for optional billing purposes.

5 Statelessness

The stateless design, which is novel in centralised orchestration architectures, is intended to ensure scalability and fault tolerance in orchestrated workflows. In the context of this work, a *pipeline* refers to a coordinated sequence of agent executions designed to accomplish a specific objective. As such, a *pipeline* is intrinsically a stateful operation, requiring the preservation of execution context across tasks to ensure correct sequencing and proper data flow.

A first approach to keeping track of *pipeline* execution state consists of adopting a stateful design. In distributed systems, this typically means that a component stores the data required for the tasks execution locally. In our centralised orchestration architecture, the orchestrator acts as the communication hub coordinating sequential task execution, while the MAS is the natural candidate for storing the pipeline state locally. Turning the MAS into a stateful component simplified orchestration and enabled faster local decisions, but it harmed scalability and fault tolerance because state coupling makes replication difficult and risks losing or desynchronising execution context on failure. These issues affect our centralised orchestration architecture design; in fact the MAS can easily become a bottleneck when handling multiple *pipeline* requests in parallel. Based on this analysis, although the statefulness of workflow execution cannot be eliminated, it is still possible to manage execution through a stateless design of the MAS. This is achieved by avoiding local state storage and instead handling state in a distributed manner. This design makes the MAS replicable and any replica can independently handle incoming requests. The following sections describe the solution adopted to address this challenge.

5.1 Workflow Preprocessing

At the core of this solution lies a preprocessing module inside the PAPI, referred to as the splitter. This module plays a crucial role in bridging the gap between a simple *workflow* definition and the operational requirements of stateless orchestration. To support a stateless design, it is essential to transform the high-level *workflow* into a collection of independent *subworkflows*, each representing an atomic task that can be executed in isolation. This decomposition lets orchestrator replicas independently execute atomic *subworkflows*, removing reliance on local state and enabling scalable, resilient orchestration. Consequently, each time a workflow is submitted through the PAPI, the splitter performs the following tasks: i) checks the *workflow* validity against platform requirements and ensures compatibility with the agents currently deployed; ii) converts the *workflow* definition from a standardised JSON format into a Python script compatible to be executed by the MAS; iii) splits the *workflow* into a portion of the overall process that can be executed independently: the *subworkflow*; iv) adapts each *subworkflow* to operate as a standalone process; v) stores the generated *subworkflow* definitions to make them available to MAS.

To clarify the execution model: each *workflow* is compiled into a single Python script composed of functions standardised by an internal orchestration library. Functions are classified as non-blocking (completed entirely within the MAS) or blocking (requiring external interaction with the AP, DH, or the user). The splitter uses blocking functions as logical boundaries to decompose the script into independent *subworkflows*. Each *subworkflow* ends by publishing a Kafka message on the communication backbone; the corresponding response triggers the next *subworkflow*. This event-driven decomposition produces loosely coupled, stateless subworkflows that can be executed independently, and the splitter reliably detects boundaries because blocking functions are predefined. The core elements resulting from this process are the *subworkflows* themselves. Each of them is implemented as a standalone Python script containing a `run` function, which is able to accept two inputs: an `OrchestratorTask` object providing execution metadata, and a dictionary containing all variables accumulated from previous subtasks. This structure enables stateless execution by explicitly passing context between subtasks without relying on local memory.

5.2 State Propagation in a Stateless Architecture

Although the MAS is designed to operate without storing execution state locally, *workflow* execution still requires passing context between tasks. Specifically, at the end of each subworkflow and just before sending the Kafka message that completes its execution, the MAS retrieves the variables generated during the subworkflow logic. This is done using Python's `inspect` module, which provides access to the call stack and execution frames. Given the known internal structure of the MAS, it is possible to navigate the stack and access the relevant frame. The local variables from that frame, representing the execution context, are extracted using the `f_locals` attribute. Once the local variables are retrieved, they are filtered to remove auxiliary or irrelevant data, ensuring that only meaningful context is preserved. The filtered variables are then Base64-encoded to ensure format compatibility. Concurrently, the MAS uses Python's inspect module to locate the frame for the running subworkflow and read its filename, which is mapped to a sequential identifier (e.g., subworkflow-1, subworkflow-2). That identifier is included in the outgoing Kafka message payload, providing the reference needed to trigger the next subworkflow. Two stateless strategies were evaluated for propagating the execution context between subworkflows, with the context stored in encoded variables. The first strategy builds on the event-driven architecture of the platform. In this approach, the state-related variables produced during workflow execution are appended directly to the messages exchanged between components. When a task is completed, the orchestrator receives the context it needs to trigger the subsequent task, without relying on local state storage. Although conceptually simple, this solution presents several drawbacks. First, it requires each Agent to return the complete accumulated state it received. This increases the risk of breaking the pipeline execution if even a single Agent fails to do so correctly. Second, the size of the messages grows linearly with the number of tasks in the workflow. This behaviour was empirically observed during testing and led to increased transmission times and noticeable delays in the later stages of execution, negatively affecting Kafka performance.

Communication efficiency is a core platform principle. For this reason, we implemented a second strategy that uses a temporary, in-memory cache, accessible to all components, to store individual task results. This avoids transmitting large payloads over Kafka and preserves Kafka's role as an event broker rather than a data store. The cache serves as a centralised and transient state store that is automatically cleared once the pipeline execution is complete, preventing unnecessary memory consumption. The overhead introduced by cache read and write operations was found to be negligible. Although this strategy introduces some additional architectural complexity, it builds on infrastructure already available in the platform: the MCS component. Its adoption enables efficient and scalable state propagation while preserving the stateless nature of the orchestrator. For these reasons, it was selected as the final implementation for our stateless orchestration model. After adopting this approach, the *workflow* execution proceeds as follows: at the end of each *subworkflow*, the variables are encoded and stored in the cache. These variables are then retrieved from the caching storage before the next *subworkflow*'s execution and passed as a dictionary to it. At the start of each *subworkflow*, the variable dictionary is decoded and the variables are reconstructed into the local context. This reconstruction lets the *subworkflow* resume execution as if it were part of a continuous, stateful pipeline while keeping the orchestrator stateless.

5.3 Error Handling

The management of execution state requires careful handling of errors and potential failures. First, the validation process performed before workflow splitting ensures that each definition is correct and free of errors, such as undefined or invalid function calls, thus preventing the splitter from processing faulty workflows. Second, all messages exchanged through the Communication Backbone are durably persisted, as the Kafka cluster is composed of three brokers with topic replication enabled. Similarly, Redis can be configured with master–replica replication, further strengthening the reliability of cached state. Finally, in case of runtime problems, for example when an agent becomes unresponsive, the MAS automatically marks the pipeline as `ERROR` once the execution timeout is exceeded. To prevent stale or inconsistent data, all cached states are assigned an expiration time, ensuring that invalid context does not persist beyond its intended lifetime.

6 Orchestration Flow

The data pipelines that are executed by the platform involve multiple communications and data exchange between several components. The reason behind this choice is to provide a complete abstraction layer to all the operational agents, by taking into consideration some technical aspects that try to optimise the entire information flow. In the next subsections, the overall dialogue between the modules will be analysed, with the aim of also drawing attention to the technical trade-offs. The starting point for executing a pipeline is to deploy all the components and all the agents that the user would like to exploit for its pipelines. A single agent can be deployed multiple times and for multiple users.

6.1 Agent Config and Upload

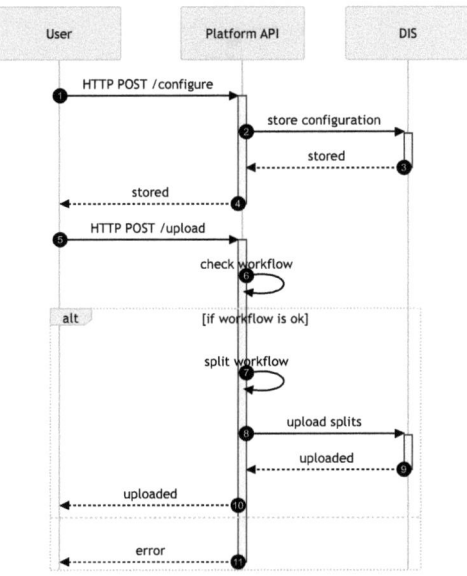

Fig. 3. The APIs for configuring an agent and uploading a workflow.

The /configure API allows users to parametrise agents at runtime. As shown in Fig. 3 (steps 1–3), these settings are stored in the DIS and later loaded by the agents before execution. This configuration phase is optional, depending on the specific use case. Workflow upload is handled through the /upload API. First, the PAPI checks the syntax of the workflow (step 6). If valid, the workflow is split into multiple subworkflows (step 7), which are uploaded and stored in the DIS (steps 8–10). This process ensures that subworkflows can later be retrieved by the orchestrator. If the validation fails, the workflow is rejected (step 11).

6.2 Pipeline Run

The execution of a pipeline begins with the /run API, as shown in Fig. 4. After the request is received (steps 1–3), the orchestrator verifies workflow consistency

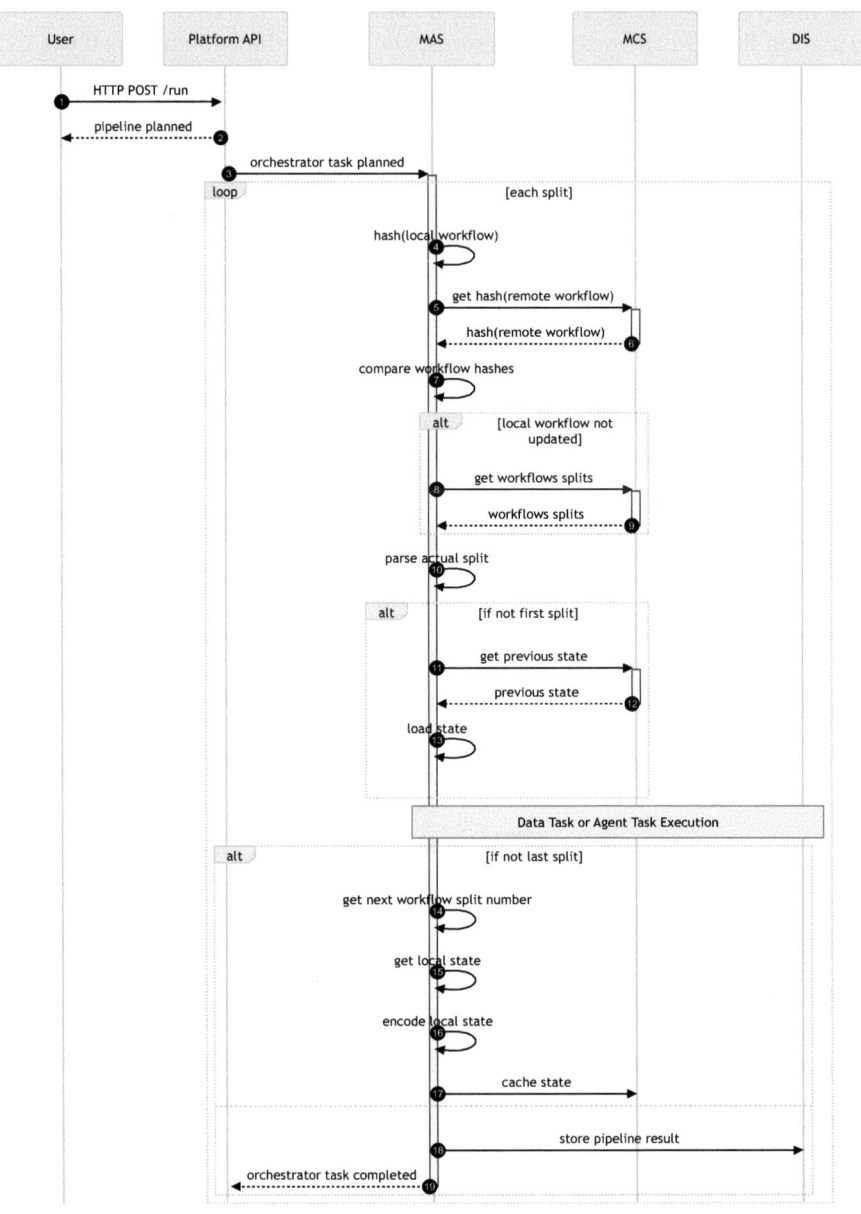

Fig. 4. Pipeline initiation and workflow management sequence. This diagram shows workflow upload verification, state restoration, and orchestration control.

by comparing the local and remote versions (steps 4–7). If necessary, updated workflow splits are retrieved from the MCS (steps 8–9). The orchestrator then parses the actual split (step 10), restores the state of the previous split from the cache and loads it (steps 11–13). Once the state is loaded, execution continues with either a **Data Task** or an **Agent Task**. If the split is not the last one, the next subworkflow number is computed and the current state is uploaded after the binary encoding (steps 14–17). Finally, the result is stored in the DIS and completion is notified to the PAPI through an **Orchestrator Task** (steps 18–19).

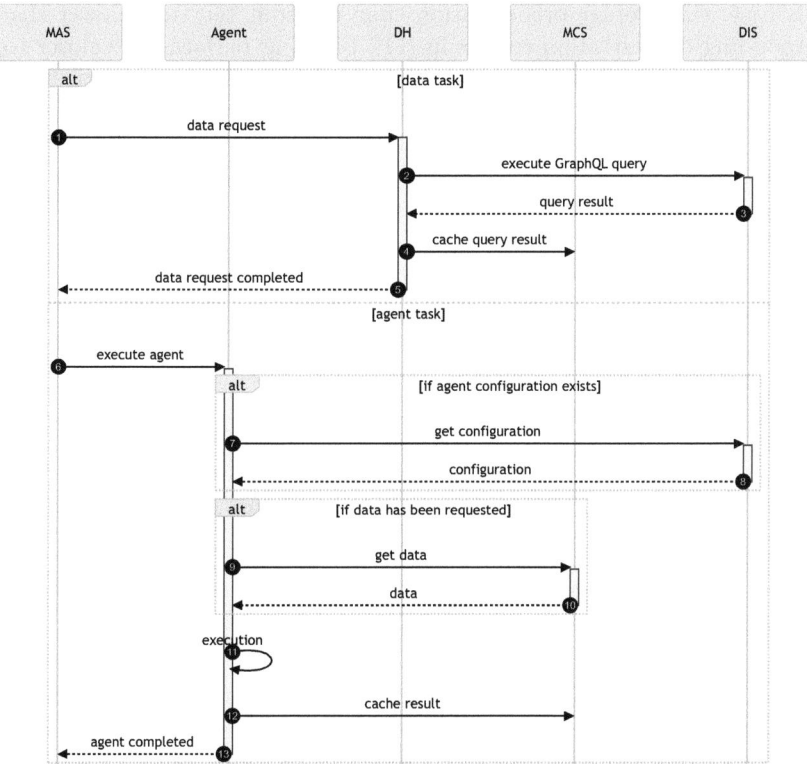

Fig. 5. Data and Agent Task execution sequence. This diagram details how the orchestrator interacts with the Data Handler and Agent Proxies during pipeline execution.

Figure 5 details these two execution paths. For data requests (steps 1–5), the orchestrator triggers the DH, which executes the query, caches the results, and signals completion through a **Data Task**. This caching solution is adopted multiple times in this work to avoid overloading the Apache Kafka CB. For instance, it could occur with the size of common historical query results. For agent execution (steps 6–13), the AP collects configuration and data (steps 7–

10), executes the assigned function (step 11), caches the output (step 12), and confirms completion (step 13) through an `Agent Task`.

7 Results

This section aims to demonstrate the viability of the platform from a performance perspective. The results presented here are based on metrics collected during execution, including latency and task duration. This execution model has been compared with the setup in [11], discussed in Sect. 2, where orchestration was reported as over 50 times slower than choreography in linear flows. In this case, the average orchestration time is about ten times lower than the corresponding orchestration results in [11], bringing performance closer to, but still below, the choreography baseline. In practice, industrial scenarios are typically characterised by a limited number of agents, making orchestration preferable due to its simpler configuration and management. Furthermore, thanks to its stateless design, our orchestrator can be replicated, achieving scalability and performance that approach those of choreography while retaining the advantages of orchestration.

Table 1. Experimental configurations by number replicas of agents (A) and orchestrators (MAS) per parallel pipeline, where p denotes pipelines, s seconds between executions, and t number of iterations.

#A	1A	2A	3A		
#MAS	1MAS	1MAS	2MAS	1MAS	2MAS	3MAS
workflow 1	1p, 4s, 3000t	2p, 4s, 1500t		3p, 4s, 1000t		
workflow 2	1p, 8s, 3000t	2p, 8s, 1500t		3p, 8s, 1000t		
workflow 3	1p, 15s, 3000t	2p, 15s, 1500t		3p, 15s, 1000t		

To assess the scalability and coordination behaviour of the orchestration framework, tests were conducted across increasing levels of pipeline complexity. Each workflow consists of a linear sequence of agent pairs, where each pair includes a simulation and an optimisation agent, serving as the unit of computational complexity. The current evaluation focuses on sequential workflows. However, the platform also supports parallel task execution, a feature that will be examined in future work. Workflow 1 and 2 comprise 2 pairs each, while Workflow 3 includes 4 pairs, leading to progressively longer execution times per trace (4, 8, and 15 s respectively). Due to the stateless design of platform components, both agents (A) and orchestrators (MAS) were deployed with up to three replicas to handle concurrent input requests. The number of MAS replicas was always less than or equal to the number of agent replicas. This led to 18 unique test configurations, as shown in Table 1, combining different replica setups for

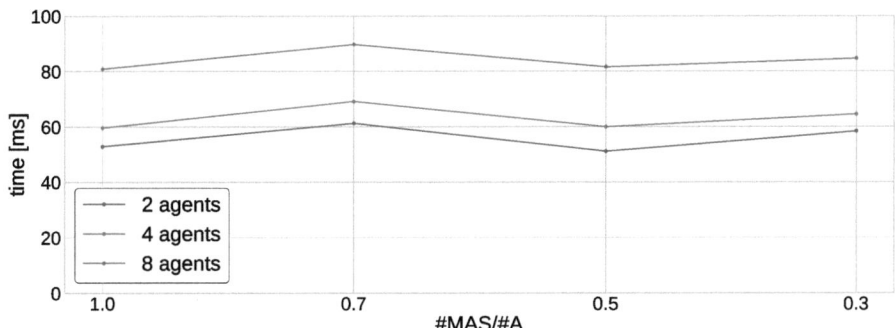

Fig. 6. Efficiency of scaling the MAS component, with multiple executions and workflows (#MAS/#A ratio). X-axis shows ratio, Y-axis orchestration time (ms).

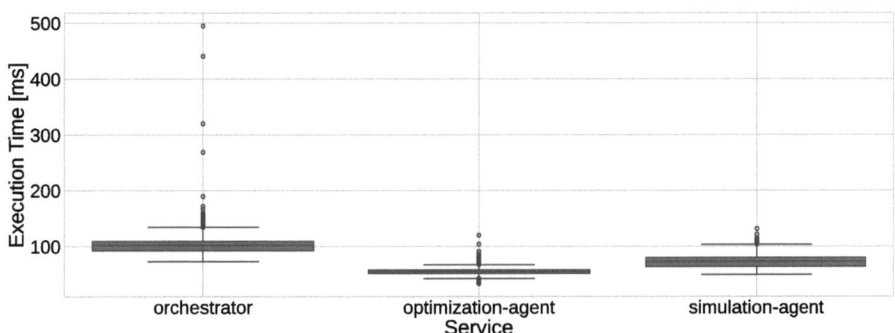

Fig. 7. Overall execution time of three main components with #MAS/#A = 1 and 2 agents in the workflow. X-axis shows kind of service, Y-axis execution time (ms).

each workflow. All tests were run on a dataset of 3,000 traces, enabling evaluation of the framework's ability to manage coordination, aggregation, and data heterogeneity under load. These tests were executed on a Kubernetes cluster deployed using the *k3s* framework. The cluster consists of 3 nodes, each implemented as a virtual machine (VM) on OpenStack, with 4 vCPUs and 10 GB of RAM. The vCPUs are provisioned from an Intel® Xeon® Gold 5318Y CPU @ 2.10 GHz. Since each vCPU corresponds to a single SMT (hyperthread) and no dedicated cores are reserved for either the host or the hypervisor, the results presented here naturally exhibit some degree of jitter.

Figure 6 illustrates the average orchestration time, as the cumulative execution time of all orchestration activities in a pipeline, across varying levels of workflow complexity (expressed as the number of agents) and different orchestrator-agent ratios (#MAS/#A). Two key observations emerge from the data. Firstly, the orchestration time does not scale linearly with the number of agents: workflows with 2, 4, and 8 agents exhibit average times around 50 ms, 60 ms, and 80 ms respectively, indicating that adding agent pairs increases overhead sublinearly, likely due to the parallel and asynchronous nature of orchestration. Second, the

#MAS/#A ratio reveals that having one orchestrator for every two agents (i.e., a ratio of 0.5) is generally sufficient to absorb the coordination load effectively. From the graph, it also emerges that the increasing communication overhead caused by adding more agents is significantly offset by the efficiency gained from executing more tasks in parallel. Performance does not significantly degrade at this ratio, suggesting that full 1:1 replication (i.e., #MAS/#A = 1.0) may be unnecessary for typical workloads. Figure 7 shows the boxplots for a single execution. The outliers are probably due to the virtual machine environment.

8 Conclusions

This paper has shown that the orchestration mechanism presented performs comparably to direct agent-to-agent communication, while offering all the benefits of orchestration. These advantages include: i) the ability to combine results from different agents, ii) logging of each execution phase, iii) monitoring of the overall process, iv) real-time data retrieval, storage, and sharing with the agents, and v) the capability to handle diverse input and output data models from individual agents, which may vary significantly. The orchestrator is also capable of handling parallel execution of subworkflows within the same pipeline, provided that the results produced by each agent are independent and do not need to be passed to one another. This capability enables the system to optimise execution time by concurrently triggering multiple agents, reducing overall latency in data-independent branches of the workflow. Finally, preliminary results confirm that parallelising these subworkflows leads to performance improvements without compromising coordination or correctness. As future steps, in order to strengthen this observation, quantitative metrics will be produced and analysed. All the tests were conducted assuming the execution of single tasks in a sequential order. The platform is, however, designed to support parallel execution of multiple tasks, and pipelines, simultaneously; evaluation of parallel, branching, and conditional workflows is left for future work. Future research will also explore integrating Agentic AI paradigms and GenAI planning modules to enhance orchestration intelligence and flexibility.

Acknowledgments. The research leading to these results has received funding from the European Community's Horizon Europe Programme under grant agreement n. 101058453.

Disclosure of Interests. The authors have no competing interests to declare that are relevant to the content of this article.

References

1. Aydin, S., Çebi, C.B.: Comparison of choreography vs orchestration based saga patterns in microservices. In: 2022 International Conference on Electrical, Computer and Energy Technologies (ICECET), pp. 1–6. IEEE (2022). https://doi.org/10.1109/ICECET55527.2022.9872665

2. Bertozzi, N., Geraci, A., Bergamasco, L., Ferrera, E., Pristeri, E., Pastrone, C.: A distributed event-orchestrated digital twin architecture for optimizing energy-intensive industries. In: Proceedings of the 10th International Conference on Internet of Things, Big Data and Security (IoTBDS 2025), pp. 337–344. INSTICC, SCITEPRESS (2025). https://doi.org/10.5220/0013364400003944
3. Chinosi, M., Trombetta, A.: BPMN: an introduction to the standard. Comput. Stand. Interfaces **34**(1), 124–134 (2012). https://doi.org/10.1016/j.csi.2011.06.002
4. Corradini, F., Marcelletti, A., Morichetta, A., Polini, A., Re, B., Tiezzi, F.: Engineering trustable choreography-based systems using blockchain. In: Proceedings of the 35th Annual ACM Symposium on Applied Computing (SAC '20), pp. 1470–1479. ACM (2020). https://doi.org/10.1145/3341105.3373988
5. Costa, B., Bachiega Jr, J., De Carvalho, L.R., Araujo, A.P.: Orchestration in fog computing: a comprehensive survey. ACM Comput. Surv. (CSUR) **55**(2) (2023). https://doi.org/10.1145/3486221
6. Garcia Lopez, P., Sanchez-Artigas, M., Paris, G., Barcelona Pons, D., Ruiz Ollobarren, A., Arroyo Pinto, D.: Comparison of FaaS orchestration systems. In: 2018 IEEE/ACM International Conference on Utility and Cloud Computing Companion (UCC Companion), pp. 148–153. IEEE, December 2018. https://doi.org/10.1109/UCC-Companion.2018.00049
7. Geiger, M., Harrer, S., Lenhard, J., Wirtz, G.: BPMN 2.0: the state of support and implementation. Future Generation Comput. Syst. **80**, 250–262 (2018). https://doi.org/10.1016/j.future.2017.01.006
8. Hahn, M., Breitenbücher, U., Kopp, O., Leymann, F.: Modeling and execution of data-aware choreographies: an overview. Comput. Sci. Res. Dev., 329–340 (2017). https://doi.org/10.1007/s00450-017-0387-y
9. Karimi, M., Barfroush, A.A.: Proposing a dynamic executive microservices architecture model for AI systems. arXiv preprint arXiv:2308.05833 (2023). https://doi.org/10.48550/arXiv.2308.05833. Accessed 28 Aug 2025
10. Pezeshkpour, P., Kandogan, E., Bhutani, N., Rahman, S., Mitchell, T., Hruschka, E.: Reasoning capacity in multi-agent systems: Limitations, challenges and human-centered solutions. arXiv preprint arXiv:2402.01108 (2024). https://doi.org/10.48550/arXiv.2402.01108, (Accessed on 28 August 2025)
11. Rudrabhatla, C.K.: Comparison of event choreography and orchestration techniques in microservice architecture. Int. J. Adv. Comput. Sci. Appl. **9**(8), 19–22 (2018). https://doi.org/10.14569/IJACSA.2018.090804
12. Singhal, N., Sakthivel, U., Raj, P.: Selection mechanism of micro-services orchestration vs. choreography. Int. J. Web Semantic Technol. (IJWesT) **10**(1), 1–13 (2019). https://doi.org/10.5121/ijwest.2019.10101
13. Svorobej, S., Bendechache, M., Griesinger, F., Domaschka, J.: Orchestration from the cloud to the edge. In: Lynn, T., Mooney, J.G., Lee, B., Endo, P.T. (eds.) The Cloud-to-Thing Continuum: Opportunities and Challenges in Cloud, Fog and Edge Computing, pp. 61–77. Palgrave Studies in Digital Business & Enabling Technologies, Palgrave Macmillan, Cham (2020). https://doi.org/10.1007/978-3-030-41110-7_4
14. Yang, Q.: Intelligent Planning: A Decomposition and Abstraction Based Approach. Artificial Intelligence. Springer, Heidelberg (2012). https://doi.org/10.1007/978-3-642-60618-2

Open Access This chapter is licensed under the terms of the Creative Commons Attribution 4.0 International License (http://creativecommons.org/licenses/by/4.0/), which permits use, sharing, adaptation, distribution and reproduction in any medium or format, as long as you give appropriate credit to the original author(s) and the source, provide a link to the Creative Commons license and indicate if changes were made.

The images or other third party material in this chapter are included in the chapter's Creative Commons license, unless indicated otherwise in a credit line to the material. If material is not included in the chapter's Creative Commons license and your intended use is not permitted by statutory regulation or exceeds the permitted use, you will need to obtain permission directly from the copyright holder.

Sustainable Solutions and Applied IoT Innovation

RAPT: AI–Powered IoT Framework for Real-Time Respiratory Disorders Monitoring and Prediction

Ritu Chauhan[1], Aarushi Mishra[1], and Dhananjay Singh[2]($\boxtimes$)

[1] Artificial Intelligence and IoT Lab, Centre for Computational Biology and Bioinformatics, Amity University, Noida, India
[2] The Pennsylvania State University, University Park, PA 16802, USA
dsingh@psu.edu

Abstract. Respiratory disorders continue to pose a significant global health concern, necessitating prompt, precise, and scalable solutions for diagnosis and monitoring. Traditional respiratory evaluations frequently rely on face-to-face examinations, restricting accessibility and the continuity of care. This work tackles the research issue of attaining continuous, distant, and precise prediction of pulmonary conditions through the use of multimodal AI combined with IoT-based e-Health systems. The main goal was to create and evaluate the Respiratory Analysis and Prediction Tool (RAPT), a system that integrates audio data and patient metadata to categorise respiratory disorders. Employing a dual-branch convolutional neural network trained on mel spectrograms and standardised clinical data (age, sex, BMI), RAPT attained a validation accuracy of 77.08% on a subset of the Respiratory Sound Database. The Gradio-powered interface facilitates real-time inference, and the tool is engineered for compatibility with IoT-enabled smart stethoscopes and wearable devices. Principal findings underscore robust efficacy in classifying COPD (precision 0.81, recall 1.00), however class imbalance constrained accuracy for less prevalent conditions such as Bronchiectasis. The paper finds that RAPT shows potential viability for personalised and remote respiratory monitoring but necessitates enhancements via bigger balanced datasets, calibration of wearable sensors, secure edge-based processing, and interoperability with 5G and Zigbee networks. This research establishes a basis for enhancing AI-driven, IoT-integrated e-Health platforms to transform respiratory health monitoring, facilitating continuous, accessible, and scalable patient care.

Keywords: Respiratory Sound Classification · IoT-Enabled Health Systems · Artificial Intelligence · Deep Learning · Convolutional Neural Network · Wearable Sensor Calibration · Telemedicine

1 Introduction

Respiratory illness, such as asthma, chronic obstructive pulmonary disease (COPD), and pneumonia, is a major global health issue, with more than 500 million individuals afflicted each year and millions of deaths, according to the World Health Organization in 2024 [1]. Respiratory illnesses tend to present themselves in the form of distinct respiratory sounds, like wheezes, crackles, and rhonchi, that are critical to diagnosis, indicating airway obstruction, alteration in lung tissue, or secretions [2]. Conventionally, auscultation with a stethoscope by experienced clinicians has been the main tool for identifying such sounds, whose quality is constrained by interpretation bias, expert skill necessity, and inability to scale to remote or underserved areas [3].

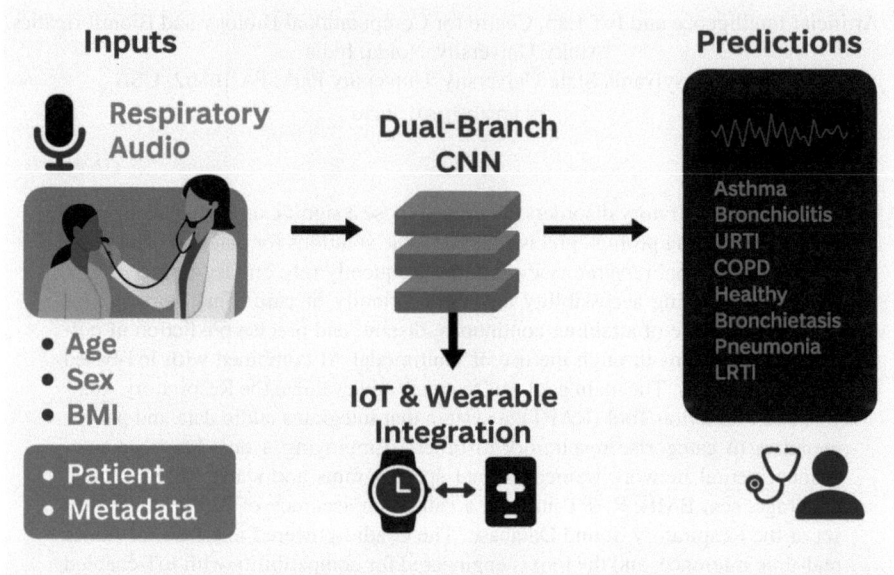

Fig. 1. IoT-enabled framework of the RAPT system for real-time classification of respiratory conditions. The pipeline starts with patient input from a smart stethoscope or wearable device. Audio and metadata are edge-processed and sent securely to the hospital server. The data is transformed into mel spectrograms and is inputted into a dual-branch convolutional neural network for classification. Real-time predictions and visualizations are presented through a Gradio-based clinician dashboard, with the ability to integrate optionally with electronic health records and cloud storage for remote monitoring in e-Health and assisted living settings.

The emergence of digital health technologies, notably e-Health and the Internet of Things (IoT), has created new opportunities for non-invasive and automated diagnosis through wearable devices and real-time processing of data to revolutionize respiratory care [4, 5]. This paper presents an AI-based respiratory analysis and prediction tool (RAPT) that diagnoses respiratory conditions from audio recordings and patient metadata, presented as a prototype for deployment over IoT platforms and wearables, thereby supporting telemedicine and assisted living solutions.

The integration of IoT in healthcare has transformed patient monitoring by making it possible to collect data continuously from dispersed devices, including smart stethoscopes, wearable sensors, and even smartphone microphones housed in smartwatches or chest patches [6–8]. These sensors, which are able to capture high-fidelity respiratory audio in real time, allow the generation of large datasets that can be analyzed using artificial intelligence (AI), especially deep learning approaches like convolutional neural networks (CNNs) [9]. IoT networks augment this potential by facilitating smooth data transfer to cloud or edge-computation platforms, where AI algorithms analyze inputs and provide actionable intelligence to patients or clinicians remotely [10]. In e-Health settings, these systems facilitate telemedicine through the minimization of in-person consultation requirements, enhanced rural access, and early identification of respiratory exacerbations, which is important for conditions like COPD that are best addressed with timely intervention [11].

Wearable technologies also enhance this capability, providing subtle monitoring of daily activity or sleep, hence facilitating assisted living for aged populations or disease management of chronic conditions [12]. Regardless of these developments, the concrete implementation of AI-based respiratory diagnostics in IoT and wearable frameworks is an area that has not been fully explored, with issues such as data privacy, device compatibility, and latency of processing [13].

The proposed RAPT meets these shortcomings by harnessing the Respiratory Sound Database, consisting of 920 recordings from 126 patients with varied conditions, to train a dual-branch CNN that handles mel spectrograms and metadata (age, sex, BMI). This not only improves classification performance but also makes the system a scalable one with a future approach to IoT integration. The tool, built with Gradio, offers a user-friendly interface for patients and clinicians alike, displaying prediction probabilities and spectrograms, and its architecture can enable data ingestion from possible IoT-enabled devices such as smart stethoscopes or smartwatches with microphones. By proposing audio and metadata communication using secure protocols (MQTT, HTTPS) to cloud or edge platforms, the system is intending to support real-time diagnostics, in line with e-Health objectives of enhancing healthcare delivery and accessibility. The potential of the prototype to continuously monitor in assisted living environments, for example, elderly care homes, further highlights its importance, providing a base for future IoT pilots in smart hospitals or home care environments. Framework of RAPT is given in Fig. 1.

The clinical and technological implications of RAPT in IoT, e-Health, and wearable platforms are immense, potentially enabling a paradigm shift in the delivery of respiratory care. The scalability of the system relies on its capability to interoperate with heterogeneous IoT devices, which calls for standardized protocols and strong edge-computing solutions to manage latency and bandwidth issues, especially in low-connectivity environments. Clinically, merging wearable streams with AI predictions may make possible tailored treatment regimens, with real-time warning systems for deteriorating conditions, improving patient outcomes for chronic disease management. But the efficacy of RAPT will have to be demonstrated in mass-scale clinical trials and IoT testbeds, resolving issues regarding battery life, data consistency, and model generalizability to diverse populations. The three chief objectives of this study are: one, to develop a robust AI model that can effectively classify respiratory disorders with combined audio and metadata inputs; two, to develop an interactive tool that supports clinical practice and telemedicine use cases; and three, to develop an IoT and wear-able deployment plan and demonstrate feasibility in simulated test cases.

This study integrates AI, IoT, and e-Health by offering a novel tool that closes the loop between machine learning and actual healthcare delivery. The contribution of this study lies in its capacity to democratize respiratory diagnosis, particularly in resource-constrained settings, utilizing the widespread availability of wearables and IoT infrastructure. However, problems related to dataset sizes, class imbalance, and validation in practical use remain, and this paper addresses these with preliminary results and a future work plan for extension. The following methodology, results, and implications sections detail the methodology, results, and implications of RAPT and discuss its scalability and performance in IoT-based e-Health environments.

2 Literature Review

There has been previous research on the application of audio-based biomarkers for respiratory disease diagnosis, against the global burden they place on health [14, 15]. There has been an attempt at examining respiratory sounds such as wheezes and crackles and cough detection in the presence of environmental noise, with significant growth in research against COVID-19 diagnosis, mobile data collection, and remote systems [14]. The Respiratory Sound Database, used here, is one of the publicly provided datasets powering these analyses [14]. Apart from this, breakthroughs in wearable and digital stethoscopes have been explored to overcome the limitations of traditional auscultation, with more emphasis on pandemics' remote care needs [22]. Non-invasive audio recording devices, like wearables and IoT, have been examined for day-to-day monitoring and potential inclusion in digital twin systems for personalized health [23], aided by the dashboard's forthcoming applications.

Previous research also explored the application of deep learning techniques in health data, particularly audio and medical data analysis. Efforts have employed Convolutional Neural Networks (CNNs) to detect Chronic Obstructive Pulmonary Disease (COPD) using features like Mel-Frequency Cepstral Coefficients (MFCC) and Mel-Spectrograms with good accuracy (e.g., 93% ICBHI score) using K-fold cross-validation [16]. Wider research into audio classification has used pre-trained CNN ensembles coupled with data augmentation to reach accuracy levels of 97% on environmental databases [17, 18]. This evidence lends credibility to the application of CNNs for extracting strong features, although the medical applications are different from environmental ones [17, 18]. Detailed reviews of deep learning in medicine point to its capacity to learn hierarchical features, which underlies the dual-branch architecture suggested by the proposed model for combining audio and metadata [19]. The work has also used recurrent neural networks like LSTMs to process medical data, although some issues like class imbalance have resulted in low recall for tasks of critical detection [16].

Other research has reviewed IoT and wearable technology in healthcare, applicable to the future implementation within a dashboard. Research into distributed-edge-computing-based IoT architectures integrated with machine learning has highlighted real-time response optimization and privacy as key areas [20]. End-to-end reviews of edge computing architectures for healthcare have highlighted low-latency applications such as vital sign monitoring, noting areas of difficulty such as data privacy and minimization [21]. The systematic study of 146 HIoT articles has classified methods as sensor-based and communication-based methods with concerns such as power control and upcoming trends such as fog computing [22]. Wearable technology has been researched for real-time monitoring with digital stethoscopes and wearables allowing remote care [15, 23], although security concerns within people-generated health data (PGHD) continue to be a problem [24]. In e-Health, telemedicine studies have demonstrated high reliability for remote imaging and chronic disease follow-up [25], suggesting potential for audio-based systems.

In spite of these developments, there remain gaps, such as limited real-world verification, small dataset sizes, and IoT interoperability issues [14, 20–22]. The tool presented here advances these endeavors by incorporating audio analysis and AI methods, suggesting a future direction for IoT and wearable integration, and seeking to promote increased accessibility and early detection in e-Health. Closing these gaps with larger datasets, strong security, and clinical trials will be essential to its success.

3 Methodology

3.1 Principle of RAPT

The respiratory analysis and prediction tool is created to deliver non-invasive AI-based classification of respiratory diseases, including asthma, COPD, and pneumonia, from audio recordings and patient metadata (age, sex, BMI). A dual-branch CNN is utilized in the system to analyze mel spectrograms of respiratory sounds and normalized metadata to deliver probabilistic predictions on eight diagnostic classes (Fig. 2).

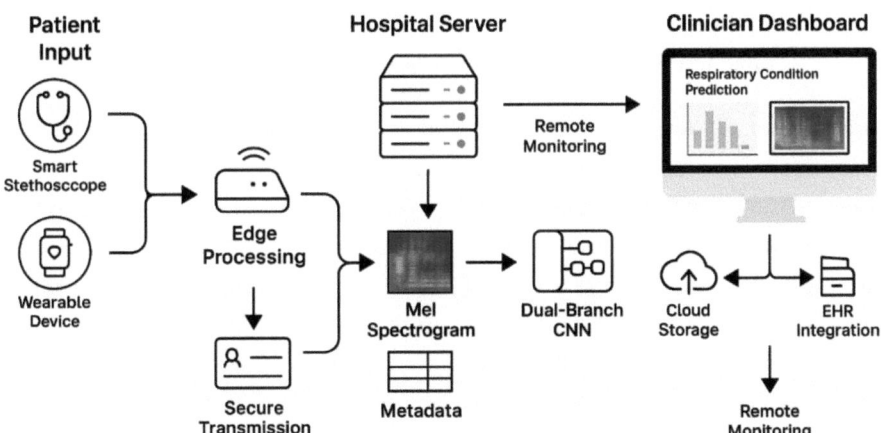

Fig. 2. Illustrates the RAPT framework, showing how respiratory audio data and patient metadata (age, sex, BMI) are processed by a dual-branch CNN to classify eight respiratory conditions. Predictions are delivered through a user-friendly smartphone interface, while IoT and wearable device integration enables scalable, real-time telemedicine and e-Health respiratory monitoring.

RAPT, deployed using Gradio, has a user-friendly interface that displays spectrograms and ordered prediction probabilities, thus being within reach for clinicians and patients. Since a prototype with a forward-thinking strategy for IoT and wearable integration, the system can be designed to communicate with devices such as smart stethoscopes or microphone wearables (e.g., smartwatches), facilitating potential real-time audio collection and sending to cloud-based or edge-based AI models using secure protocols such as MQTT. This would enable real-time monitoring in telemedicine systems and assisted living settings, such as nursing homes, consistent with e-Health paradigms through enabling scalable, remote diagnosis. The modular nature of the prototype would enable flexibility in fitting into IoT ecosystems with possible use on edge devices such as Raspberry Pi for low-latency computation in environments with limited resources.

3.2 Dataset Description and Preprocessing

The Respiratory Sound Database, derived from an open repository, contains 920 annotated audio recordings of 126 patients, 5.5 h in duration, and capturing 6898 respiratory cycles (1864 with crackles, 886 with wheezes, 506 with both) [26]. The dataset also contains diagnoses for eight conditions (Healthy, Asthma, COPD, URTI, Bronchiectasis, Pneumonia, Bronchiolitis, LRTI) and metadata (age, sex, BMI, child weight, child height), captured using devices such as 3M Littmann and WelchAllyn stethoscopes, descriptive statistics of the dataset is provided in Table 1. Audio preprocessing consisted of loading the files at 22,050 Hz for 3 s using Librosa, extracting mel spectrograms (64 mel bins, 4,000 Hz top frequency, 1,024 FFT window, 256 hop length), and performing augmentation (time stretching with 0.9–1.1 rate, ± 1 semitone pitch shift, 0.2% noise injection) to increase robustness. Spectrograms were trimmed or padded to a fixed size (64, 259) with -80 dB silence.

Table 1. Descriptive statistics of the patient dataset used in the study

Metric	Patient_id	Age	Sex	Adult_BMI	Child_Weight	Child_Height	Condition
Count	126	126	126	126	126	126	126
Mean	163.500	43.127	0.642	27.275	17.286	101.217	3.642
STD	36.517	32.115	0.497	4.134	10.496	17.803	1.556
Min	101	0.250	0.000	16.500	7.140	64.000	0.000
25%	132.250	4.250	0.000	26.223	15.100	99.500	3.000
50%	163.500	60.000	1.000	27.400	15.100	99.500	3.000
75%	194.750	70.750	1.000	28.250	15.100	99.500	4.000
Max	226.000	93	2.000	53.500	80.000	183.000	7.000

Metadata were parsed and missing values were imputed with median strategy, while age was normalized through StandardScaler; sex was one-hot encoded as F, M and yielded a 7-dimensional vector because there was an extra NA column. EDA consisted of frequency distribution plots of diagnoses (indicating COPD predominance with 17/26 validation samples), feature correlation heatmaps (e.g., age vs. BMI), box plots of features (e.g., BMI vs. condition, showing range for COPD patients), and descriptive statistics (e.g., median age 27.4, BMI 27.4), verifying data quality and balance concerns. Patient-level aggregation averaged spectrograms for each patient, resulting in 126 samples (100 training, 26 validation).

3.3 Model Architecture

The classification model consists of a dual-branch architecture using TensorFlow that combines audio and metadata inputs to make strong predictions for eight respiratory diseases. The audio branch handles mel spectrograms (shape $64 \times 259 \times 1$) using two Conv2D layers: one with 32 filters (3×3 kernel, ReLU activation, same padding), followed by batch normalization and 2×2 max pooling; another with 64 filters, batch normalization, and max pooling. Global average pooling downcasts spatial dimensions to a 64-dimensional vector. The metadata branch calculates a 7-dimensional vector (age, BMI, child weight, child height, F, M, NA) via a dense layer of 64 units (ReLU activation) with batch normalization. Outputs of both branches are combined into a 128-dimensional vector, fed into a dense layer of 128 units (ReLU activation), and finally into a last dense layer of 8 units (softmax activation) to obtain class probabilities. The 37,512 parameter (37,192 trainable) model is trained using Adam (initial learning rate of 0.001) and categorical cross-entropy loss, accuracy and AUC metrics being used. Overfitting is reduced through batch normalization and global pooling to make it appropriate for noisy real-world data.

3.4 Model Training Pipeline

The pipeline for training was tuned for efficiency and stability, tackling the class imbalance of the dataset as well as memory. The 126 patient samples were divided into 100 train and 26 validation to avoid data leakage. A custom DataGenerator, subclassing TensorFlow's Sequence, produced batches of 8, which included audio inputs ($8 \times 64 \times 259 \times 1$), metadata inputs ($8 \times 7$), and one-hot label ($8 \times 8$) with shuffled patient IDs per epoch. Class weights were computed to assign greater weight to sparsely represented classes (e.g., Bronchiectasis, LRTI), according to the class distribution of the training set. Maximum training of 100 epochs with the following callbacks: EarlyStopping (patience = 20, monitoring validation AUC, restoring best model weights), ReduceLROnPlateau (factor = 0.5, patience = 5, min 1e-6), and ModelCheckpoint (best model saved to 'best_model.keras'). The maximum validation accuracy of 77.08% and AUC of 0.9147 were achieved in epoch 31, even though AUC calculations failed in final test due to non-normalized probabilities. Memory optimization included TensorFlow memory growth configuration and batch processing, using 13 batches to train and 4 batches for validation per epoch.

3.5 RAPT Implementation

RAPT, which is implemented with Gradio, offers an interactive interface for real-time respiratory condition classification, taking in WAV audio files, age, sex (Male/Female), and BMI as inputs. Audio is processed to mel spectrograms using the same training pipeline, padded to 64 × 259 × 1, whereas metadata is normalized and padded to fit the 7-dimensional training input, using child weight and height median values from the preprocessed dataset. The learned model outputs probabilities for eight conditions as a sorted, color-coded HTML list (e.g., red for > 50% probabilities), and the input audio's spectrogram. The user interface, with response time less than 2 s, has clinical usability and provides example inputs (e.g., 'example_healthy.wav') with no caching to avoid mistakes. The design of RAPT facilitates telemedicine accessibility, with outputs being integrable into electronic health records, and increases its function as an e-Health tool.

3.6 IoT and Wearable Integration

RAPT is designed as a future prototype with an IoT and wearable integration approach for enabling real-time monitoring of respiratory in e-Health and assistive living. Audio is possibly attainable through IoT-enabled hardware like smart stethoscopes (e.g., Eko) or microphones on wearables (e.g., smartwatches) at 22,050 Hz, which is supported by the model's input requirement. Information can be transmitted to cloud or edge AI models using secure transport protocols like MQTT for lightweight messaging or HTTPS for secure transfers, along with compression (e.g., WAV to MP3) to minimize bandwidth. Real-time processing can be enhanced via cloud computing or refined TensorFlow Lite models on edge devices like Raspberry Pi, enabling low-latency predictions in distant settings. The framework may be integrated with telemedicine platforms to send predictions to doctors or patients, and provide for ongoing monitoring in care-assisted living settings (e.g., geriatric care) by sampling audio regularly during daily routines. Data privacy challenges may be met through encryption, battery life through intermittent sampling, and connectivity through offline buffering, with real-world experiments required to confirm feasibility for smart hospitals and home care.

4 Results

4.1 Dataset Preprocessing Outcomes

Preprocessing of the Respiratory Sound Database, including 920 recordings from 126 patients, provided a solid dataset for model evaluation and training. Audio recordings were converted into 64 mel bin mel spectrograms with a 3-s length to have a uniform shape of 64 × 259 pixels, as observed in a representative spectrogram in Fig. 3. This spectrogram created using Librosa displayed distinct frequency patterns for breathing cycles, featuring tightly clustered vertical ridges indicating unusual breath sounds. Techniques like time stretching, pitch shifting, and adding Gaussian noise were utilized to improve the variability and resilience of the data.

Metadata preprocessing included missing value imputation via median strategy and normalization of features. Descriptive statistics of the 126 patients were reported as a mean age of 43.13 years (SD 32.12), mean BMI of 27.28 kg/m^2 (SD 4.13), and male predominance (64.29%), represented in Table 1. Child weight and height had medians of 15.1 kg and 99.5 cm, respectively, consistent with the pediatric subset of the dataset. In Fig. 4, EDA, performed through Pandas and Seaborn, comprised frequency distribution plots, which illustrated that there was a skewed distribution with COPD being the most common condition among validation samples (17/26), while categories like Bronchiectasis and Healthy were underrepresented.

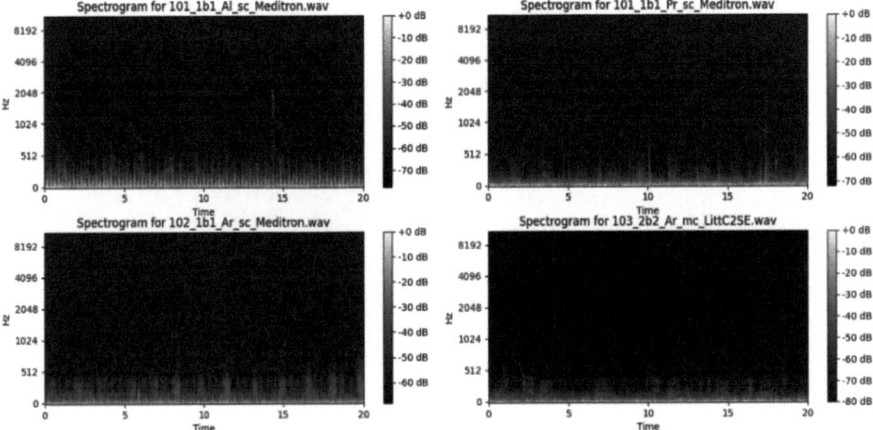

Fig. 3. Mel spectrogram from audio preprocessing.

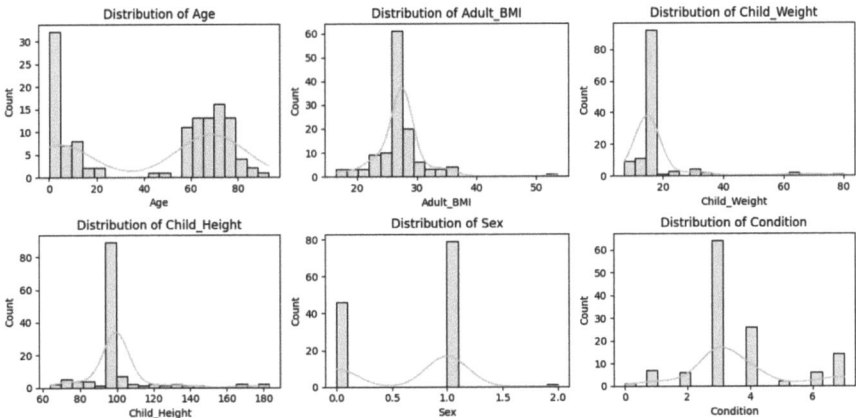

Fig. 4. Frequency distribution of numerical features across the dataset.

A feature correlation heatmap, depicted in Fig. 5, demonstrated weak to moderate linear correlations among variables, including a slight positive correlation between age and BMI ($r \approx 0.2$), indicating feature independence and lowering the likelihood of multicollinearity. These analyses confirmed the overall data quality and diversity while noting significant class imbalance, which was subsequently addressed using focal loss and class weighting in model training.

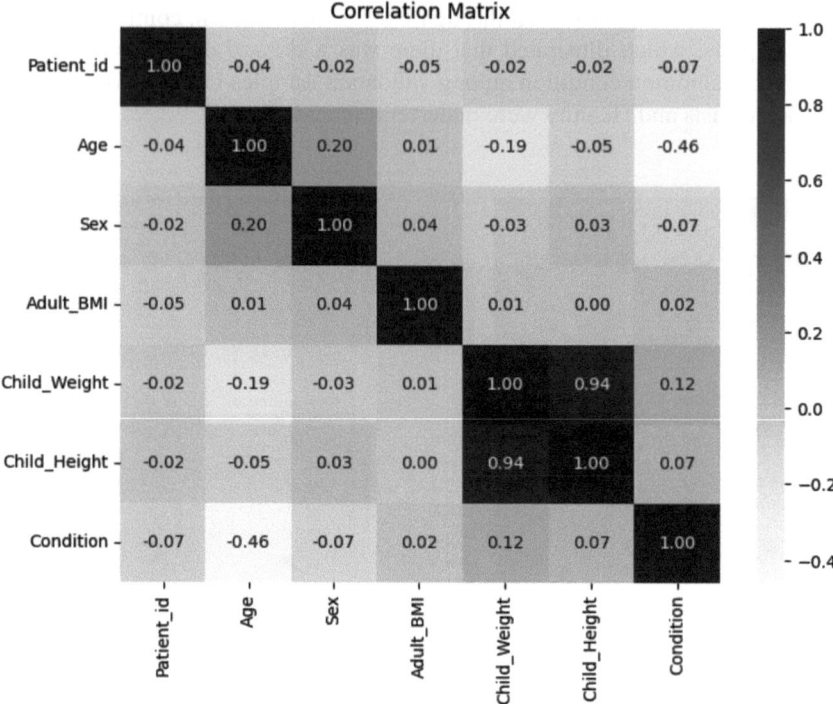

Fig. 5. Heatmap showing the correlation matrix between numerical patient features.

4.2 Model Performance Results

The two-branch CNN model, combining mel spectrograms and metadata, achieved a validation accuracy of 77.08% on the 26-patient validation set, as reported in the classification report. The model did well for COPD, with precision 1.00, recall 0.94, and F1-score 0.97, indicating its predominance, whereas URTI and Bronchiectasis had average findings as presented in Table 2.

Table 2. Classification report of the model on the validation set

	Precision	Recall	F1-score	Support
Healthy	0.0	0.00	0.00	3
COPD	1.00	0.94	0.97	17
UTRI	1.00	0.50	0.67	4
Bronchiectasis	0.29	1.00	0.44	2
Accuracy			0.77	26
Macro Avg.	0.57	0.61	0.52	26
Weighted Avg	0.83	0.77	0.77	26

Healthy had zero recall and precision because of small samples pointing to the effect of imbalance. The weighted average F1-score was 0.77, indicating fair overall performance. The confusion matrix, as shown in Fig. 6 also verified sixteen true positives for COPD, two true positive each for URTI and Bronchiectasis, and no misclassifications, consistent with the report's high COPD recall.

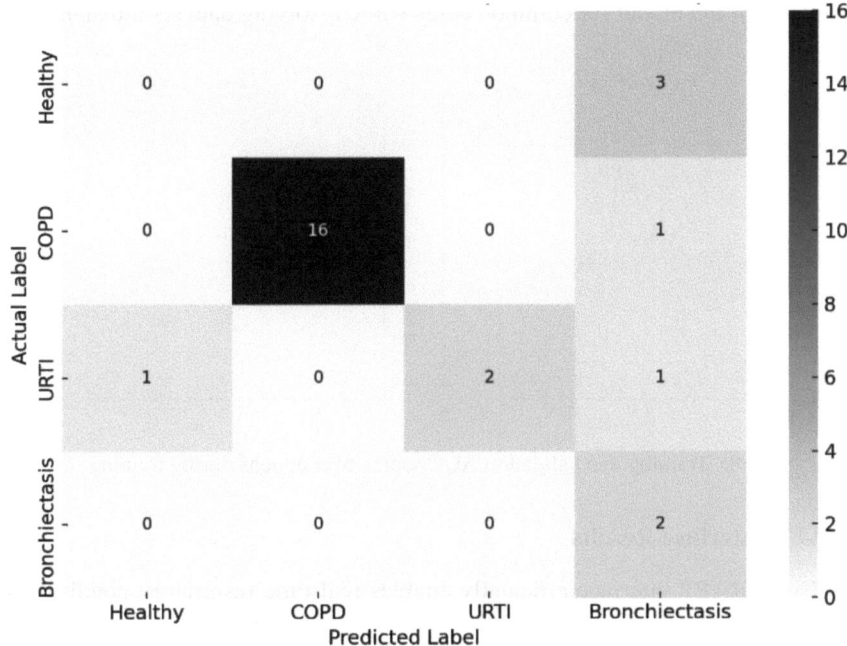

Fig. 6. Confusion matrix for the final model evaluated on the test set

Training dynamics, plotted in accuracy and loss curves in Fig. 7, revealed training accuracy up to 0.7 and validation accuracy leveling off at 0.4–0.5 across 50 epochs with validation loss declining sharply to 5–10 before oscillating.

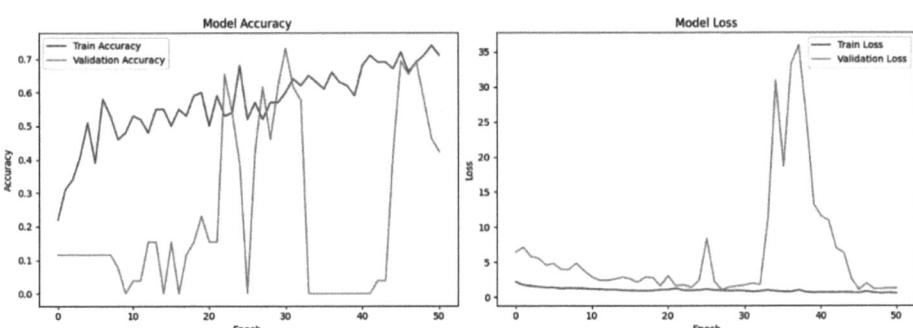

Fig. 7. Model accuracy and loss trends over 50 training epochs for training and validation sets

Figure 8 plots the trends of model accuracy and loss during 50 epochs of training, where training accuracy approaches about 0.7, whereas validation accuracy plates around 0.4–0.5, indicating the effect of an imbalanced class distribution, specifically the dominance of COPD (17/26 validation samples). Validation loss has an early steep drop to 5–10, oscillations, which are due to fast convergence to salient features and self-regulation from class weighing and sparse validation data. Such trends, controlled by optimization methods such as Adam with adaptive learning rate decay, underscore the resilience of the model for common cases while resolving data set imbalances.

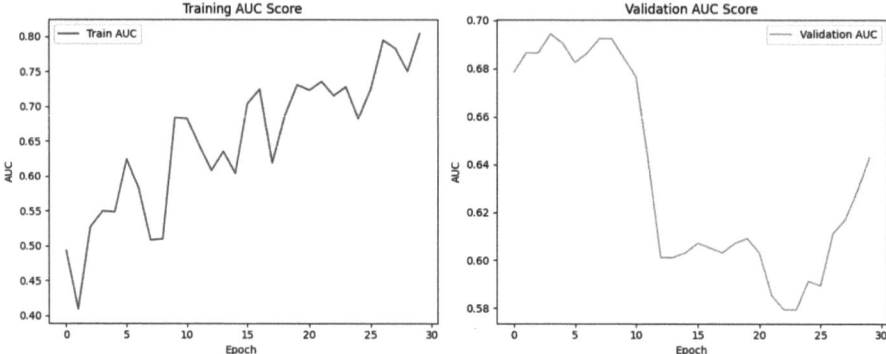

Fig. 8. Training and validation AUC scores over epochs during training.

4.3 RAPT Interface Results

Gradio-based RAPT interface efficiently enables real-time respiratory condition classification interactively and intuitively for clinicians and patients. Users upload WAV audio files of the respiratory data and input metadata (sex, age, BMI), initiating backend processing to create mel spectrograms and condition predictions in a 2-s turnaround. A sample of the input and output interface in Fig. 9 clearly illustrates an easily defined field for the upload of audio along with entry of patient information. This input design aims for ease of use and readability, facilitating usage in both clinical and non-clinical environments. The output interface shows the Mel spectrogram generated, which visualizes patterns of time-frequency intensity commonly examined by clinicians to review the occurrence of wheezes, crackles, or airway obstructions. The Prediction Results panel shows a ranked list of the probabilities of each condition, side by side. In the illustrative example, LRTI is expected with absolute certainty (100%), while the remaining conditions (for example, Asthma, COPD, Bronchiectasis) are correctly held back to 0%, justifying the model's specificity. Probabilities are red coded for immediate interpretability conditions with >50% confidence as an aid to decision-making for non-expert users.

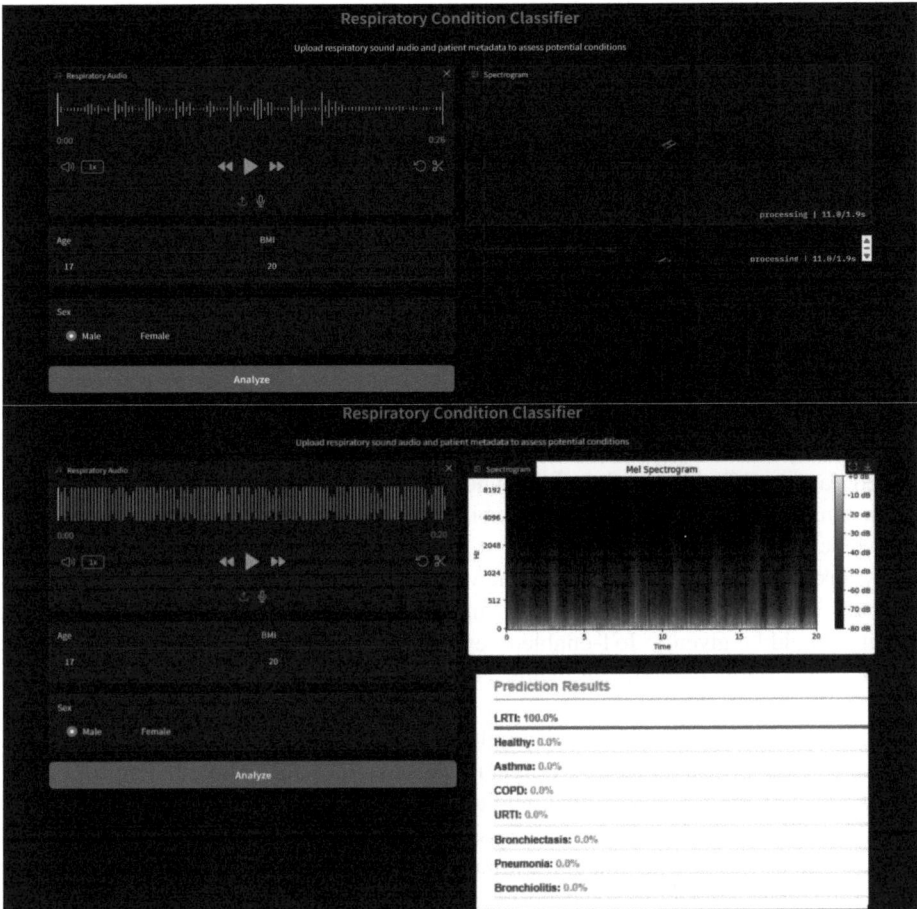

Fig. 9. The input and output interface of RAPT. Output interface displays the spectrogram visualization and condition prediction

This frontend-backend integration with Gradio and deep learning replicates the model's 77.08% validation accuracy, according to the classification metrics. It places RAPT at par as a deployable e-Health system, particularly for telemedicine and IoT settings. Supported by HTML-styled outputs and negligible latency, the system can be seamlessly integrated into wearable systems, such as smart stethoscopes or health devices, for remote respiratory monitoring and decision support automation.

4.4 Future Approach for IoT Integration

The design of RAPT as a prototype can be incorporated into IoT platforms and wearables, including smart stethoscopes or microphone-enabled smartwatches, to provide real-time monitoring of respiration as represented in Fig. 1. Audio at 22,050 Hz can be transformed into mel spectrograms, data transmission being made through MQTT or HTTPS to cloud or edge platforms, which can support continuous monitoring (e.g., 5 min) for elderly care. Edge processing with TensorFlow Lite-optimized models on devices like Raspberry Pi would be able to maintain prediction accuracy, while data privacy may be ensured by encryption. The approach suggests scalability in home care or smart hospitals, though experimentation in the real world will be needed to overcome connectivity challenges, battery life, and calibration issues.

5 Conclusion

This study revealed that the Respiratory Analysis and Prediction Tool (RAPT), a multimodal AI-based framework, can proficiently diagnose respiratory disorders by utilizing audio features and patient metadata, with a validation accuracy of 77.08% with robust COPD detection capabilities. The study suggests that AI-enhanced, IoT-integrated e-Health systems can revolutionize respiratory health monitoring by facili-tating personalized, remote, and continuous care. The study encountered difficulties, such as a limited dataset (126 cases), class imbalance, and an absence of real-world validation, which constrained the model's generalizability and accuracy for infrequent illnesses like Bronchiectasis. Based on these findings, it is advisable for healthcare professionals and policymakers to include AI-driven frameworks such as RAPT into telemedicine practices, underpinned by standardized data standards and secure edge-based infrastructures. Priority should be given to IoT-enabled devices, such smart stethoscopes and wearable sensors, for scalable deployment, while assuring data privacy and interoperability with 5G or Zigbee networks.

Future research must rectify existing limitations by acquiring larger, balanced, and diverse datasets, investigating lightweight models like MobileNet for efficient edge computing, and performing longitudinal field trials in smart hospitals. Moreover, enhancing the calibration of wearable sensors and assessing user acceptance in practical environments would be essential to guarantee reliability and confidence in clinical applications. These guidelines will facilitate the closure of current disparities and expedite the secure, efficient, and equitable implementation of AI-driven respiratory e-Health solutions.

Disclosure of Interests. The authors have no competing interests to declare that are relevant to the content of this article.

References

1. World Health Organization. Chronic respiratory diseases: more than 80 million affected and many more undiagnosed, warns new WHO and European Respiratory Society report, WHO/Europe, Jun. 12, 2025. https://www.who.int/europe/news/item/12-06-2025-chronic-respiratory-diseases--more-than-80-million-affected-and-many-more-undiagnosed--warns-new-who-and-european-respiratory-society-report
2. Singh, S.: Respiratory symptoms and signs. Medicine (Baltimore) **48**(4), 225–233 (2020). https://doi.org/10.1016/j.mpmed.2020.01.001
3. Seah, J.J., Zhao, J., Wang, D.Y., Lee, H.P.: Review on the advancements of stethoscope types in chest auscultation. Diagnostics **13**(9), 1545 (2023). https://doi.org/10.3390/diagnostics13091545

4. Kaur, H., Atif, M., Chauhan, R.: An Internet of Healthcare Things (IoHT)-based healthcare monitoring system. In: Mohanty, M.N., Das, S. (eds.) Advances in Intelligent Computing and Communication, LNNS, vol. 109, pp. 475–482. Singapore: Springer Singapore (2020). https://doi.org/10.1007/978-981-15-2774-6_56
5. Yafi, E., Chauhan, R., Sharma, A., Zuhairi, M.F.: Integrated empowered AI and IoT approach for heart prediction. In: 2024 18th International Conference on Ubiquitous Information Management and Communication (IMCOM), pp. 1–7. IEEE, Kuala Lumpur, Malaysia (2024). https://doi.org/10.1109/IMCOM60618.2024.10418366
6. Ahmad, R.U.S., Khan, M.S., Hilal, M.E., Khan, B., Zhang, Y., Khoo, B.L.: Advancements in wearable heart sounds devices for the monitoring of cardiovascular diseases. SmartMat **6**(1), e1311 (2025). https://doi.org/10.1002/smm2.1311
7. Chauhan, R., Singh, D.: Predictive analytics for stress management in nursing: a machine learning approach using wearable IoT devices. In: Singh, D., Van 'T Klooster, J.-W., Tiwary, U. S. (eds.) Intelligent Human Computer Interaction, LNCS, vol. 15557, pp. 60–75. Springer Nature Switzerland, Cham (2025). https://doi.org/10.1007/978-3-031-88705-5_6
8. Li, C., Wang, J., Wang, S., Zhang, Y.: A review of IoT applications in healthcare. Neurocomputing **565**, 127017 (2024). https://doi.org/10.1016/j.neucom.2023.127017
9. Phatak, A.A., Wieland, F.-G., Vempala, K., Volkmar, F., Memmert, D.: Artificial intelligence based body sensor network framework—narrative review: proposing an end-to-end framework using wearable sensors, real-time location systems and artificial intelligence/machine learning algorithms for data collection, data mining and knowledge discovery in sports and healthcare. Sports Med. - Open **7**(1), 79 (2021). https://doi.org/10.1186/s40798-021-00372-0
10. Rafique, W., Qi, L., Yaqoob, I., Imran, M., Rasool, R.U., Dou, W.: Complementing IoT services through software defined networking and edge computing: a comprehensive survey. IEEE Commun. Surv. Tutor. **22**(3), 1761–1804 (2020). https://doi.org/10.1109/COMST.2020.2997475
11. George, A.S., George, A.H.: Telemedicine: a new way to provide healthcare. Partners Univ. Int. Innovat. J. **1**(3), 98–129 (2023)
12. Rattanawiboomsom, V., Talpur, S.R.: Enhancing health monitoring and active aging in the elderly population: a study on wearable technology and technology-assisted care. Int. J. Online Biomed. Eng. **19**(11) (2023). https://doi.org/10.3991/ijoe.v19i11.41929
13. Putra, K.T., et al.: A review on the application of internet of medical things in wearable personal health monitoring: a cloud-edge artificial intelligence approach. IEEE Access **12**, 21437–21452 (2024). https://doi.org/10.1109/ACCESS.2024.3358827
14. Kapetanidis, P., et al.: Respiratory diseases diagnosis using audio analysis and artificial intelligence: a systematic review. Sensors **24**(4), 1173 (2024). https://doi.org/10.3390/s24041173
15. Troncoso, Á., Ortega, J.A., Seepold, R., Madrid, N.M.: Non-invasive devices for respiratory sound monitoring. Procedia Comput. Sci. **192**, 3040–3048 (2021). https://doi.org/10.1016/j.procs.2021.09.076
16. Kim, Y., et al.: Evolution of the stethoscope: advances with the adoption of machine learning and development of wearable devices. Tuberc. Respir. Dis. **86**(4), 251–263 (2023). https://doi.org/10.4046/trd.2023.0065
17. Srivastava, A., Jain, S., Miranda, R., Patil, S., Pandya, S., Kotecha, K.: Deep learning based respiratory sound analysis for detection of chronic obstructive pulmonary disease. PeerJ Comput. Sci. **7**, e369 (2021). https://doi.org/10.7717/peerj-cs.369
18. Nanni, L., Maguolo, G., Brahnam, S., Paci, M.: An ensemble of convolutional neural networks for audio classification. Appl. Sci. **11**(13), 5796 (2021). https://doi.org/10.3390/app11135796
19. Nanni, L., Costa, Y.M.G., Aguiar, R.L., Mangolin, R.B., Brahnam, S., Silla, C.N.: Ensemble of convolutional neural networks to improve animal audio classification. EURASIP J. Audio Speech Music Process. **2020**(1), 8 (2020). https://doi.org/10.1186/s13636-020-00175-3

20. Yen-Wei, C.: Deep learning in healthcare: paradigms and applications. Intelligent Systems Reference Library (2020)
21. Alnaim, A.K., Alwakeel, A.M.: Machine-learning-based IoT–edge computing healthcare solutions. Electronics **12**(4), 1027 (2023). https://doi.org/10.3390/electronics12041027
22. Hartmann, M., Hashmi, U.S., Imran, A.: Edge computing in smart health care systems: review, challenges, and research directions. Trans. Emerg. Telecommun. Technol. **33**(3), e3710 (2022). https://doi.org/10.1002/ett.3710
23. Haghi Kashani, M., Madanipour, M., Nikravan, M., Asghari, P., Mahdipour, E.: A systematic review of IoT in healthcare: applications, techniques, and trends. J. Netw. Comput. Appl. **192**, 103164 (2021). https://doi.org/10.1016/j.jnca.2021.103164
24. Pergolizzi, J.V., LeQuang, J.A., El-Tallawy, S., Varrassi, G.: What Clinicians should tell their patients about wearable devices and data privacy (2024). https://doi.org/10.20944/preprints202409.1428.v1
25. Maita, K.C., et al.: Imaging evaluated remotely through telemedicine as a reliable alternative for accurate diagnosis: a systematic review. Health Technol. **13**(3), 347–364 (2023). https://doi.org/10.1007/s12553-023-00745-3
26. Rocha, B.M., et al.: A respiratory sound database for the development of automated classification. In: Maglaveras, N., Chouvarda, I., De Carvalho, P. (eds.) Precision Medicine Powered by pHealth and Connected Health, IFMBE Proceedings, vol. 66, pp. 33–37. Springer Singapore, Singapore (2018). https://doi.org/10.1007/978-981-10-7419-6_6

Open Access This chapter is licensed under the terms of the Creative Commons Attribution 4.0 International License (http://creativecommons.org/licenses/by/4.0/), which permits use, sharing, adaptation, distribution and reproduction in any medium or format, as long as you give appropriate credit to the original author(s) and the source, provide a link to the Creative Commons license and indicate if changes were made.

The images or other third party material in this chapter are included in the chapter's Creative Commons license, unless indicated otherwise in a credit line to the material. If material is not included in the chapter's Creative Commons license and your intended use is not permitted by statutory regulation or exceeds the permitted use, you will need to obtain permission directly from the copyright holder.

Digital Product Passport as Digital Carrier for Information of Life Cycle Assessment: A Feasibility Study of Solvolysis on Composite Recycling for Wind Turbines Blades

Christina Tsitsiva and Michail J. Beliatis(✉)

Aarhus University, 7400 Herning, Denmark
mibel@btech.au.dk

Abstract. The study incorporates the concept of a digital product passport as a data carrier for tracking the end-of-life of wind turbine blades across the processes value chain, integrating circular economy principles into the business model. Evaluation of a novel chemical recycling process for wind turbine blades using Life Cycle Assessment (LCA) and Life Cycle Cost Analysis (LCCA) suggests that the chemical recycling process has significantly reduces environmental impacts when compared to landfill disposal and offers potential economic benefits. However, scalability and material recovery efficiency remain key factors for widespread adoption.

Keywords: Wind turbine blades · Life Cycle Assessment · Life Cycle Cost Analysis · Digital Product Passport · Sustainability · Circular Economy · Eco-design

1 Introduction

As wind energy demand increases, the end-of-life management of wind turbine blades remains a challenge due to their composite structure. The shift to renewable energy sources is crucial for managing climate change. Wind energy, in particular, plays an important role and the blades of the wind turbines are key components. Blades contain a complex material mix that includes glass fiber, thermosets like epoxy, thermoplastics like PVC, PET and PU, balsa wood, adhesives, and coatings. Due to this complex mix, blades pose a challenge for material recycling since it's difficult to separate them [1]. Epoxy resin, favored for its strength and lightweight properties, enhances the durability and structural integrity of these blades. However, the difficulty in separating these materials complicates recycling efforts.

Currently, end-of-life wind turbines can be recovered up to 80% on average through standard recycling routes [2]. Although the blades and composite materials remain the main challenge for the wind sector, they can be recycled through thermal, chemical, or mechanical processes. Thermal recycling includes heating the blades at high temperatures, mechanical recycling reduces blades to particles, and chemical recycling employs solvents to break down the components for reuse. Despite efficiency and research regarding the chemical process, industrial-scale implementation remains limited due to high energy requirements and high-cost catalysts for the chemical process. A limited number of, recent studies show that recovering bisphenol A (BPA) from epoxy resins can boost recycling efficiency while minimizing energy consumption. [3] leading to increased value for the end product.

This preliminary theoretical study investigates the economic and environmental impacts of solvolysis as a method for separating epoxy from fibers in wind turbine blade recycling on an industrial scale. Upcoming EU legislation is expected to force companies to disclose information related to recycled materials and the recyclability ratio of products through the Digital Product Passport (DPP) [4]. Therefore, blockchain technology could improve the DPP by incorporating raw material data, in a secure, trusted and decentralized way, supporting further the recycling efforts from digital perspective [5]. Conclusively, the research aims to explore the feasibility and development of a business model framework prioritizing material recovery, recycling, and sustainability, while focusing on waste hierarchy and reusing activities.

2 Literature Review

2.1 Wind Turbine Blade Waste Management

The wind industry is a sustainable, cost-effective and clean energy source. EU's commitment to reducing greenhouse gas emissions by up to 95% in the coming decades, imposes that wind turbines are likely to play an important part in the future energy supply [6].

The life expectancy of wind turbines being around 25 years, along with the forecasted high demand for wind turbines, estimates that 325 kilotons of wind turbine blades-of-life in EU the forthcoming decade. For that reason, there is ongoing research based on waste management prioritizing circularity and more specifically on blades due to their complexity. Most wind turbines contain steel, copper and aluminum meaning that almost 94% can be recycled [7].

The disposal of decommissioned turbine blades involves significant environmental problems, such as the generation of microplastics and pollution of soil and water sources and economic challenges for financing the operations for the decommission [6]. Traditional waste management methods, such as landfilling and incineration, are considered unsustainable due to their harmful effect on the environment and public health. As a result, there is an urgent need for creative and sustainable waste management techniques that are consistent with the concepts of a circular economy, which prioritizes resource recovery, recycling, and waste reduction [8].

Current research identifies many approaches that handle end-of-Life wind turbine blades, including mechanical recycling, thermal processing, and repurposing for new uses [6]. However, technological, economic, and legal barriers frequently impede the application of these approaches. To successfully address the rising waste issue, it's essential to develop composite materials that are recyclable at the end of their life cycle [7].

2.2 Recycling

Effective recycling solutions are required to reduce environmental impact, recover valuable resources, and promote sustainability in the sector [9]. To solve these issues, a variety of recycling techniques have been developed, each with its own set of advantages and limits.

Mechanical

Mechanical recycling is one of the most popular ways of treating end-of-life wind turbine blades. Typically, mechanical shredders are used to break down composite materials into smaller particles. One of the key benefits of mechanical recycling is that it uses less energy than other recycling technologies, making it an economically viable choice. However, mechanical recycling has its limits. The quality and consistency of the recovered materials might vary because the procedure might not sufficiently separate the various components of the composite materials. Furthermore, recycled materials may have lower mechanical qualities than virgin materials, limiting their usage in vital uses [10].

Chemical

Chemical recycling known also as solvolysis is a cutting-edge way of recycling end-of-life wind turbine blades, using chemical processes to break down composite materials into each of their components. This method not only improves the quality of the recovered components, but it also allows for the regeneration of the resin, which can then be reused in new composite compositions. However, chemical recycling can be more complicated and energy-intensive than mechanical or thermal techniques. The need for specialized equipment and well-defined procedures might result in greater operational expenses, which might limit adoption. Furthermore, the quality and consistency of recovered materials are strongly reliant on the exact chemical procedures used, demanding stringent quality control techniques [10].

Thermal

Thermal recycling refers to a variety of methods, including pyrolysis and oxidation. This recycling process comes along with increased energy consumption and operating expenses. The amount of energy required to heat the materials can be large, raising questions about the process's overall sustainability. Furthermore, the possible losses during heat degradation might restrict the recovery of useful materials [10].

An alternative to pyrolysis is co-processing. Co-processing is a waste management technique that uses combustion to convert waste products into fuel and raw material. This process promotes a circular economy since it helps to conserve energy and resources in other material production processes, reducing industry dependency on raw and virgin materials that are consequently decreasing at an unsustainable rate [9].

In conclusion, the effectiveness of wind turbine blade recycling methods varies based on their impact over time. Mechanical recycling is a viable short-term option because of its low energy consumption and economic feasibility, but the low grade of recovered materials restricts its long-term sustainability. Thermal recycling, which includes pyrolysis and oxidation, is a medium-term solution that balances material recovery with increased energy needs and operating costs. However, worries about sustainability and efficiency might prevent broad implementation. Chemical recycling, despite its complexity and energy-intensive nature, provides a long-term solution by recovering high-quality materials which makes it a potential technique for a circular economy. As the wind energy sector strives to increase sustainability, advances in recycling technology and regulatory guidance will be critical in optimizing these processes for broad adoption.

2.3 Legislation Framework for Wind Turbine Blades

The legislative framework governing wind turbine blade waste management in the European Union is primarily shaped by the European Waste Framework Directive (2008/98/EC). The framework establishes a waste hierarchy that prioritizes prevention, reuse, recycling, recovery, and disposal. This directive serves as the basis for national waste policy, pushing the wind sector to adopt circular economic concepts. Despite the directive's broad scope, it lacks precise recycling objectives for composite components, like wind turbine blades. In response to increasing decommissioning volumes, different member states are implementing specific measures. For example, France is proposing regulating measures that would mandate wind turbine recycling, which might create an example for other countries [11]. The Circular economy action plan [12] strengthens this transition by encouraging the recycling of complex materials and discourages landfill disposal, indicating a steady regulatory move toward more strict sustainability criteria.

A key policy instrument that could enhance wind turbine blade waste management is Extended Producer Responsibility (EPR), which shifts waste management responsibilities from governments to manufacturers. EPR has been widely applied in industries such as electronics and packaging, resulting in increased recycling rates and lower waste disposal costs [13]. However, its application to wind turbines is hampered by regulatory exemptions. Wind turbine blades are designated as Large-Scale Fixed Installations (LSFIs), so they are excluded from the Waste Electrical and Electronic Equipment (WEEE) Directive. [11] legal gap restricts industry-wide responsibility for blade end-of-life management, leaving disposal decisions primarily up to project owners rather than manufacturers. Furthermore, the lack of a developed market for secondary materials generated by decommissioned blades, along with high recycling prices, creates financial and logistical obstacles.

Emerging regulatory frameworks, such as the Digital Product Passport (DPP), present opportunities to improve traceability and transparency in wind turbine blade waste management. The DPP is intended to give extensive product life cycle information, such as material composition, repairability, and disposal alternatives [4]. By including standardized "track and trace" data, the DPP connects with programs like the Sustainable Products Initiative (SPI) and the Eco-design Directive, which aim to set minimum sustainability criteria across industries [14]. Nevertheless, obstacles remain in integrating the DPP into current legal frameworks, especially in terms of identifying data ownership, maintaining accessibility between EU member states, and aligning with business demands. If effectively implemented, the DPP might lead to more efficient recycling, material recovery, and regulatory compliance, complementing the EU's overall goal of a circular economy [5]. The advancement of these legal frameworks will help shape the future of wind turbine blade waste management, bridge present policy gaps, and ensure long-term end-of-life solutions for renewable energy infrastructure.

3 Methods

3.1 Life Cycle Assessment

Life Cycle Assessment (LCA) is a systematic process for assessing the environmental implications of a product at all phases of its life, from raw material extraction to manufacture, use, and disposal. The purpose of LCA is to offer a complete perspective of a product or service's environmental elements and prospective consequences, assisting in the identification of chances for improvement and supporting decision-making processes.

Furthermore, LCA is regulated by international standards like ISO 14040 and ISO 14044, which give a structure and procedures for performing LCA research. The Life Cycle Analysis (LCA) process begins with Goal and Scope Definition, defining the study's objectives and analysis boundaries. The Life Cycle Inventory (LCI) collects data on energy, material inputs, and environmental discharges. Then, the Life Cycle Impact Assessment (LCIA) assesses environmental implications, including global warming potential and resource depletion. Lastly, the Interpretation phase analyzes the data to make informed judgments, detect key concerns, and provide suggestions [15].

The functional unit for this study is defined as (1) wind turbine blade. System boundaries include raw material recovery, chemical processing, energy consumption, and disposal alternatives. The impact categories assessed include CO_2 emissions, water consumption, energy demand, and waste generation. Data sources include OpenLCA software with secondary data from industry reports and databases. A sensitivity analysis is conducted to assess the variability of results under different energy mix scenarios.

3.2 Data Collection

This study investigates the solvolysis process for wind turbine blade recycling by utilizing a combination of primary and secondary data sources. Primary data was collected through a semi-structured interview with an environmental protection professional from a wind turbine manufacturer, offering critical insights into technological feasibility, operational challenges, and regulatory landscape of blade recycling. The expert's perspective provided a practical understanding of the industry's efforts to minimize environmental impact, enhance energy efficiency, and integrate circular economy principles. Secondary data was obtained from academic literature, industry reports, and regulatory documents, covering key topics such as life cycle assessment (LCA) methodologies, life cycle cost analysis (LCCA), and legislative frameworks governing wind turbine end-of-life management.

4 Results

4.1 Environmental Impact Analysis

Conducting an LCA provides key visualizations, such as model graphs and Sankey diagrams, which offer a comprehensive overview of material and energy flows. The model graph that is generated in the software illustrates the interconnected processes involved in each end-of-life of the blades, allowing them to track inputs, outputs, and environmental burdens systematically (Fig. 1).

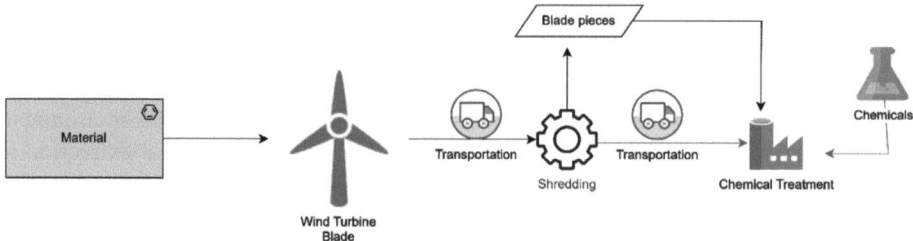

Fig. 1. Value Chain Model graph of Treated Composite Blade

Sankey diagrams, on the other hand, visually represent the magnitude of environmental impacts across different stages, highlighting spots where resource consumption and emissions are most significant. These tools enable a clearer interpretation of complex data, supporting a more transparent comparison between solvolysis and landfill disposal of wind turbine blades (Fig. 2).

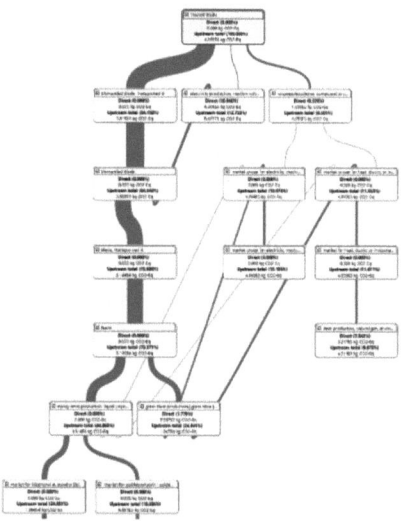

Fig. 2. Sankey diagram from Treated Blade showing the impact of each process/component in the recycling process

The environmental implications of wind turbine blade disposal are a crucial factor in evaluating end-of-life management strategies. Figure 3, compares the environmental implications of two end-of-life management options for wind turbine blades: landfill disposal and solvolysis treatment (treated blade) with EF v3.1 impact assessment method. The evaluation assesses a variety of environmental effect categories, such as water usage, particulate matter generation, material resource depletion, and several toxicity and climate change indicators [16].

The results highlight the environmental burdens across multiple impact categories: *Climate Change and Acidification*

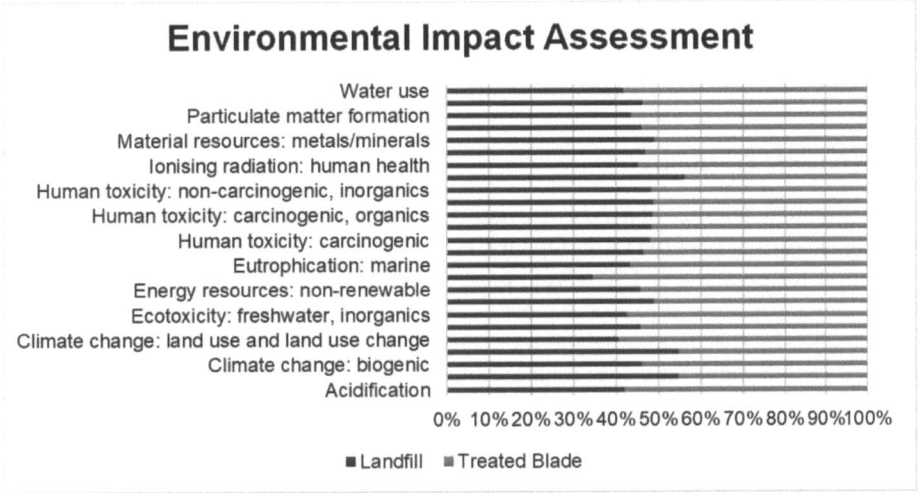

Fig. 3. Environmental Impact Assessment with EF v3.1 as impact assessment method

Solvolysis has more climate change related impacts, including land usage and biogenic contributions, than landfill disposal. The increased energy consumption in solvolysis, along with emissions from chemical inputs, results in a larger carbon footprint. Even though, the overall climate change is less than the landfill's. Moreover, acidification, caused by acidic component emissions, is also increased in solvolysis, indicating the chemical processes involved in breaking down the blade materials.

Water Usage and Energy Resources

Solvolysis treatment uses more water and drains energy resources than landfill disposal. This is mostly owing to the energy-intensive procedures and the usage of enormous amounts of chemicals like isopropanol and organophosphorus compounds in solvolysis. The increased energy demand results in a larger reliance on nonrenewable energy supplies, increasing environmental impact.

Human Health and Toxicity

Solvolysis treatment has a greater influence on human toxicity (both carcinogenic and non-carcinogenic) and ionizing radiation. These effects are caused by the creation and use of chemicals throughout the treatment procedure. In contrast, the landfill process has very low toxicity consequences, because it involves less active processing. However, landfills produce pollutants such as methane and carbon dioxide, which contribute to both local environmental damage and worldwide climate change.

Material Resources and Eutrophication

Material shortage (metals and minerals) is more severe in the solvolysis treatment due to the increased inputs required for chemical processes and the necessity for specialized catalysts. Solvolysis has a stronger eutrophication impact in marine habitats, due to probable chemical emissions and waste byproducts.

Overall, while solvolysis provides a path toward resource recovery and circularity, it also could impose high environmental costs in certain impact categories, particularly in water use, toxicity, and acidification. These factors must be weighed against the long-term sustainability challenges associated with landfill disposal to determine the most effective strategy for wind turbine blade end-of-life management.

4.2 Economic Feasibility

The economic feasibility of solvolysis versus landfill disposal could be easily evaluated by using the software for the LCA. OpenLCA and SimaPro software tools provide a solid foundation for conducting Life Cycle Cost Analysis (LCCA) by including economic evaluations into Life Cycle Assessment (LCA) either manually or through databases. These tools allow users to attribute monetary values to various phases of a product's life cycle. So, it makes it easier to assess the financial viability of various end-of-life techniques by including cost data directly into environmental models. For instance, OpenLCA, allows for scenario-based cost comparisons, which aid in assessing differences in processing costs, transportation distances, and regulatory fees. This method allows decision-makers to examine not just direct economic costs but also prospective income from material recovery and indirect expenses, such as carbon fees or landfill limitations.

While short-term costs may favor landfill disposal, this option often neglects long-term liabilities and the lost value of wasted materials. On the contrary, solvolysis presents an opportunity to recover valuable materials like glass fiber and epoxy resin, which can generate revenue and help offset operational expenses. This makes it a more economically sound long-term solution. Preliminary theoretical investigations suggest that solvolysis is a more cost-effective option than landfilling. However, further verification of pilot studies is necessary to confirm these findings.

For the early calculation of the total profit per process the following equation is adopted:

$$\text{Total Profit} = \text{Outputs} - \text{Inputs}$$

Total Profit (Landfill) = - License cost for the landfill
Total Profit (Treatment) = Revenue from Recovered Materials - (Cost of Materials + Energy consumption)

While solvolysis offers a compelling economic case for material recovery, its profitability relies on energy price swings and catalyst reuse efficiency which requires future pilot investigations to validate these hypotheses under industrial settings. The high cost of specialized infrastructure, such as industrial autoclaves, prevents broad adoption till nowadays. Furthermore, scaling up the process presents new technological challenges, such as maintaining consistent catalyst performance, reducing operational waste, and optimizing energy use. Addressing these concerns through targeted investments and collaborative initiatives is critical to ensuring that solvolysis delivers economic and environmental benefits in industrial settings.

In conclusion, the preliminary theoretical economic feasibility analysis showed that solvolysis is a more cost-effective choice for wind turbine blade disposal than landfill. While solvolysis needs a significant initial investment due to high input costs, especially for specialist catalysts and chemicals, its capacity to recover valuable materials such as bisphenol A (BPA), glass fiber (GF), and metals makes up for these costs. In contrast, landfill disposal results in a direct financial loss due to waste processing and disposal expenses, with no possibility of material recovery.

4.3 DPP Enabled Business Ecosystem

Addressing the end-of-life management of wind turbine blades necessitates an integrated recycling ecosystem. This ecosystem involves collaboration among manufacturers, material suppliers, recycling operators, transportation companies, and regulatory bodies. The Digital Product Passport (DPP) could enhance traceability and compliance across the entire value chain network, promoting a digital circular economy and facilitating transparency in material recovery and reuse. Figure 4 illustrates a circular business ecosystem in which different players work together to ensure circularity in wind turbine blade materials. This ecosystem is made up of multiple major entities, each with a critical function aiding material recovery, regulatory compliance, and digital traceability.

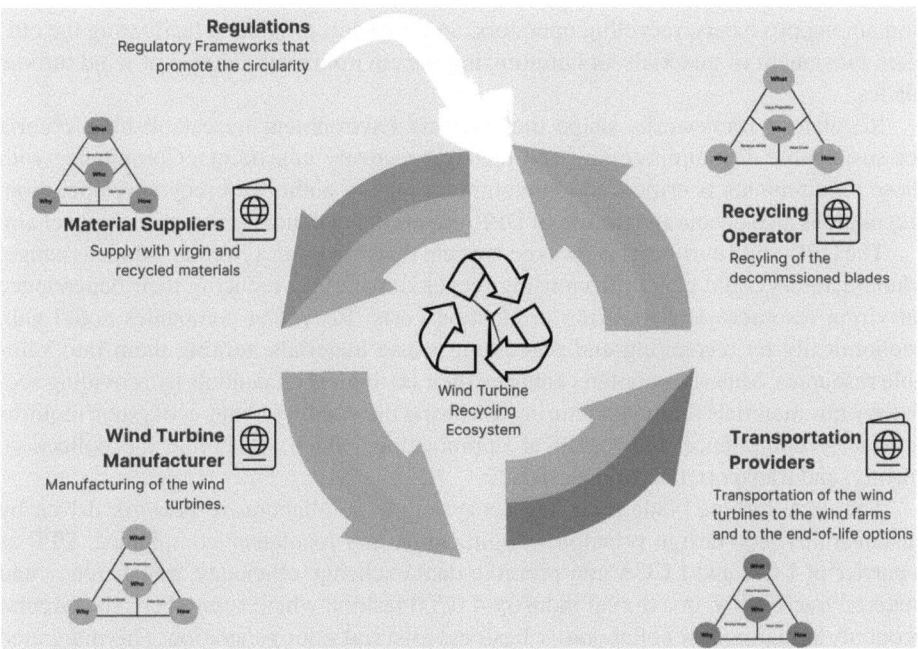

Fig. 4. Business Ecosystem of blade recycling value chain for solvolysis

The wind turbine manufacturer is crucial to the business ecosystem since it manufactures wind turbine blades. The DPP enhances value creation since it enables traceability throughout the product's lifespan, allowing stakeholders to view material composition, maintenance history, and end-of-life treatments. For that reason, the manufacturer should work with material suppliers, transportation companies, and recycling companies to create an efficient recycling process. Furthermore, the introduction of the recycling service could create new revenue streams from the sale or reuse of materials in the wind turbine.

Material suppliers contribute to the upstream supply chain by providing raw materials for wind turbine blade manufacturing. They could also contribute to the circular economy by reusing materials from decommissioned blades to create novel products, positioning themselves as sustainable suppliers. Furthermore, the integration of DPP allows for effective material tracking, assuring compliance with sustainability norms.

The recycling operator oversees the processing decommissioned blades using modern recycling procedures like solvolysis to recover precious components like fiberglass and epoxy. With DPP's data help, it is possible to optimize the recycling process by offering insights into material qualities and previous changes. Recovered materials could be then returned to material providers or the wind turbine manufacturer, completing the circularity loop.

Transportation companies support the transportation of blades from wind farms to recycling facilities. Given the logistical problems of carrying big wind turbine blades, effective route planning and specialized equipment to dismantle blades are required to save costs and environmental impact. Details provided in the DPP can improve cooperation among producers, recycling operators, and transport providers, facilitating the efficient movement of materials and minimizing the environmental impact of wind turbine blades.

Regulatory frameworks shape the business environment by establishing criteria for sustainable decommissioning and material recovery constraints. Compliance with these requirements is critical to maintaining a legally authorized recycling operation. Regulations also encourage the use of DPP, increasing openness across the supply chain.

The DPP could nurture a business ecosystem that thrives on a circular value exchange. Manufacturers could benefit from using recycled materials, reducing their dependence on virgin resources and lowering production costs. Recycling companies could gain economically by recovering and processing waste materials, turning them into valuable resources. Material suppliers enhance their ecological credentials by providing secondary raw materials for manufacturing. Ensuring the viability of this ecosystem requires regulatory compliance and logistical optimization, which streamline the collection, sorting, and transportation of materials.

The wind turbine blade recycling ecosystem is a collaborative network driven by sustainability, eco-design principles, digitization, and regulatory compliance. DPP as a carrier of LCA and LCCA can promote data exchange efficiency, transparency and material traceability, in a digital industry 4.0/5.0 fashion which strengthens the circular economy human centric collaborative business trust and value generation. This integrated system highlights how industry stakeholders may collaborate to reduce waste, reduce emissions, and improve resource efficiency in the wind energy sector.

4.3.1 Implementation Roadmap of DPP

The effective execution of a Digital Product Passport for wind turbines needs a methodical, strategic 3 phase plan that transitions from theoretical comprehension to a concrete, functional system.

The first phase is the Foundation Phase outlines the strategic framework for the initiative, including defining the blueprint of data points for end-of-life processes like solvolysis and data point requirements for LCA/LCCA. Furthermore, the design of a comprehensive data governance architecture, specifying ownership and security measures. This phase ensures that collaboration between key stakeholders, such as manufacturers, material suppliers, and recycling operators, must be established to enable interoperability and data exchange in strategized industry 4.0 fashion. After the groundwork, phase 2 takes place with a Pilot system development for early validation before a full-scale launch. This prototype enables stakeholders to test functionality and identify issues in real-world settings, which helps to improve the data model, streamline operations, and guarantee user-friendliness. Finally, the Integration phase takes place where the DPP is integrated with enterprise systems like Enterprise Resource Planning (ERP) and Warehouse Management (WHM) via Application Programming Interfaces (APIs). This phase leads to automated exchange of data flow at production ready systems, guaranteeing uniform application of the DPP and emphasizes continues improvements, including the integration of advanced elements such as carbon footprint metrics with IoT or the implementation of blockchain technology for improved security and traceability (Fig. 5).

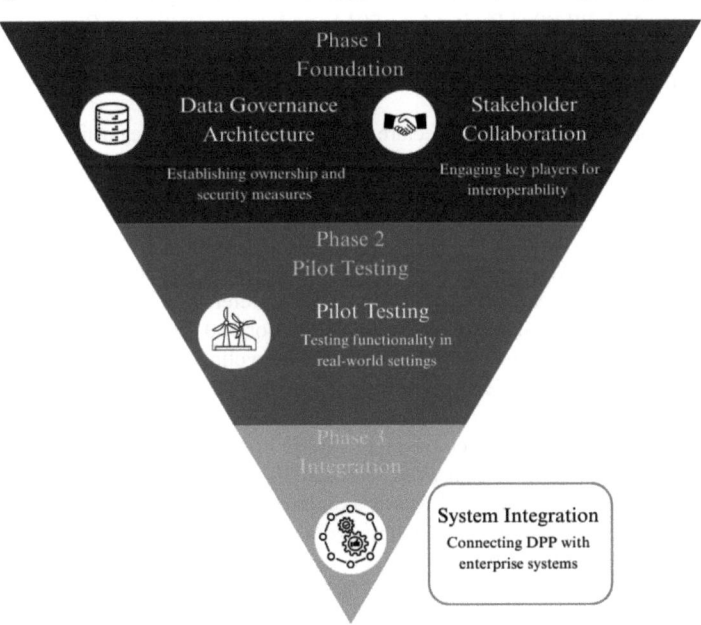

Fig. 5. Implementation Process of Digital Product Passport

5 Conclusions

In conclusion, this study shows that using a chemical recycling method for wind turbine blades could significantly improve sustainability by reducing waste and helping with material recovery. The theoretical LCA findings show that solvolysis has a lower environmental effect than landfilling, especially in terms of decreased emissions and resource saving. Similarly, the LCCA demonstrates that, while the initial investment in recycling infrastructure is significant, the long-term benefits, such as recovered materials and regulatory incentives, could make it a feasible economic option, but further verification using pilot studies are needed. Furthermore, DPP could serve as a digital carrier of LCA and LCCA to increase traceability, optimize end-of-life decision-making, and allow for a more circular approach to blade management.

Transitioning from a linear disposal method to a circular system raises both obstacles and possibilities in terms of business models. While the addition of recycling services necessitates strategic changes to logistics, relationships, and financial planning, it also creates new income streams through material recovery, reuse, and compliance with new environmental standards. Companies who embrace this transformation will gain a competitive advantage in the renewable energy market by proving their dedication to sustainability and innovation.

Acknowledgments. Michail J. Beliatis acknowledge the Interreg Baltic Sea Region Project: GlassCircle Index No.: PIFS004 Project No.: S014 and Interreg Baltic Sea Region Project: CompositeCircle Index No.: PIFC3.2079 Project No.: C072 for supporting partially the developed artifact at Aarhus University.

Disclosure of Interests. The authors declare no conflicts of interest regarding this manuscript.

References

1. Mattsson, C., André, A., Juntikka, M., Tränkle, T., Sott, R.: Chemical recycling of end-of-life wind turbine blades by solvolysis/HTL. In: IOP Conference Series: Materials Science and Engineering. IOP Publishing Ltd. (2020). https://doi.org/10.1088/1757-899X/942/1/012013
2. Khalid, M.Y., Arif, Z.U., Hossain, M., Umer, R.: Recycling of wind turbine blades through modern recycling technologies: a road to zero waste (2023). https://doi.org/10.1016/j.ref.2023.02.001
3. Ahrens, A., et al.: Catalytic disconnection of C-O bonds in epoxy resins and composites. Nature **617**, 730–737 (2023). https://doi.org/10.1038/s41586-023-05944-6
4. Adisorn, T., Tholen, L., Götz, T.: Towards a digital product passport fit for contributing to a circular economy. Energies **14** (2021). https://doi.org/10.3390/en14082289
5. Singh, P., et al.: Blockchain for economy of scale in wind industry: a demo case. Presented at the January 1 (2023). https://doi.org/10.1007/978-3-031-20936-9_14
6. Tayebi, S.T., Sambucci, M., Valente, M.: Waste management of wind turbine blades: a comprehensive review on available recycling technologies with a focus on overcoming potential environmental hazards caused by microplastic production (2024). https://doi.org/10.3390/su16114517
7. Woo, S.M., Whale, J.: A mini-review of end-of-life management of wind turbines: current practices and closing the circular economy gap. Waste Manag. Res. **40**, 1730–1744 (2022). https://doi.org/10.1177/0734242X221105434

8. Deeney, P., et al.: End-of-life alternatives for wind turbine blades: sustainability indices based on the UN sustainable development goals. Resour. Conserv. Recycl. **171**, 105642 (2021). https://doi.org/10.1016/j.resconrec.2021.105642
9. Paulsen, E.B., Enevoldsen, P.: A multidisciplinary review of recycling methods for end-of-life wind turbine blades. Energies **14** (2021). https://doi.org/10.3390/en14144247
10. Jani, H.K., Singh Kachwaha, S., Nagababu, G., Das, A.: A brief review on recycling and reuse of wind turbine blade materials. Mater. Today Proc. **62**, 7124–7130 (2022). https://doi.org/10.1016/j.matpr.2022.02.049
11. Majewski, P., Florin, N., Jit, J., Stewart, R.A.: End-of-life policy considerations for wind turbine blades. Renew. Sustain. Energy Rev. **164** (2022). https://doi.org/10.1016/j.rser.2022.112538
12. Circular economy action plan. https://environment.ec.europa.eu/strategy/circular-economy-action-plan_en. Accessed 19 June 2025
13. Extended Producer Responsibility. https://www.oecd.org/en/publications/extended-producer-responsibility_67587b0b-en.html. Accessed 19 June 2025
14. The European Green Deal, Brussels (2024)
15. Horne, R.E., Grant, T., Verghese, K.L.: Prospects for life cycle assessment development and practice in the quest for sustainable consumption
16. Zampori, L., Pant, R.: Suggestions for updating the organization environmental footprint (OEF) method. Publications Office of the European Union (2019)

Open Access This chapter is licensed under the terms of the Creative Commons Attribution 4.0 International License (http://creativecommons.org/licenses/by/4.0/), which permits use, sharing, adaptation, distribution and reproduction in any medium or format, as long as you give appropriate credit to the original author(s) and the source, provide a link to the Creative Commons license and indicate if changes were made.

The images or other third party material in this chapter are included in the chapter's Creative Commons license, unless indicated otherwise in a credit line to the material. If material is not included in the chapter's Creative Commons license and your intended use is not permitted by statutory regulation or exceeds the permitted use, you will need to obtain permission directly from the copyright holder.

Bridging ESG and Capability Maturity: A Case-Based Artefact for Industrial Organisations

Lasse Cenholt and Mirko Presser(✉)

Department of Business Development and Technology, Aarhus University, 7400 Herning, Denmark
mirko.presser@btech.au.dk

Abstract. Organisations preparing for the European Sustainability Reporting Standards (ESRS) mandate often lack a clear, structured approach to assess the maturity of their Environmental, Social, and Governance (ESG) capabilities.

This study develops an ESG-specific Capability Maturity Model (ESG-CMM) through a six-step Design Science Research process. A systematic review distilled 104 descriptors from 15 capability maturity model articles. Open coding, cross-checked against ESRS topics, condensed these into 13 parameters arranged within five transversal clusters and five classic CMM levels.

The artefact was demonstrated in a mid-sized manufacturing firm via seven semi-structured interviews. Evidence-backed ratings placed the organisation at Level 2.0 ("Repeatable"), highlighting strengths in governance policy and gaps in analytics capability and integrated risk management.

The ESG-CMM artifact thus bridges a documented academic gap by translating a software-derived maturity logic to ESG practice, offering managers a self-assessment tool that is both regulator aligned and implementation oriented. While the single-case design limits generalisability, the research sets a foundation for multi-site validation and for sensor-driven, real-time scoring extensions.

Keywords: Environmental-Social-Governance · Capability-Maturity Model · Industrial Internet of Things · Sustainability · Analytics

1 Introduction

Environmental, Social, and Governance (ESG) refers to a company's environmental impact, its responsibilities to employees, stakeholders, the supply chain, and the wider community, as well as the governance structures that ensure ethical and accountable decision-making. The focus on ESG has grown significantly with the introduction of new legislative requirements, particularly the European Sustainability Reporting Standards (ESRS) under the Corporate Sustainability Reporting Directive (CSRD). This legislation requires a higher level of transparency and accountability in ESG reporting than was previously expected in sustainability disclosures [1–3].

The Capability Maturity Model (CMM) was originally developed in the late 1980s at Carnegie Mellon University's Software Engineering Institute. Initiated by Watts S. Humphrey, the model was created as a systematic approach to improve the quality and predictability of large software projects. Humphrey introduced the concept of process maturity, suggesting that organisational capabilities evolve through five qualitative stages: Initial, Repeatable, Defined, Managed, and Optimising [4].

Building on this foundation, Mark Paulk and his colleagues formalised the framework in 1991 with the release of Capability Maturity Model for Software, Version 1.0 [5]. This version introduced the five familiar maturity levels and linked each stage to specific Key Process Areas (KPAs), giving managers a clear path for assessment and improvement. Feedback from industry use led to refinements in Version 1.1 (1993), which clarified documentation, added performance metrics, and emphasised integration across project activities [6]. The model continued to evolve through later versions, becoming SW-CMM in 1997 and eventually CMM Integration (CMMI). CMMI releases between 2000 and 2018 expanded the model's scope beyond software, incorporating agility, risk management, and enterprise-wide flexibility [7–9]. This progression illustrates how a model originally designed for software development became a widely applicable framework for continuous improvement across diverse organisational settings.

CMM provides a structured framework for evaluating capabilities through its five maturity levels. These levels help organisations understand their current state and identify opportunities for growth. Its adaptability makes CMM suitable for applications beyond software, including emerging areas such as ESG practices. Despite its broad potential, no specific CMM currently exists for ESG. This research aims to fill that gap by developing an ESG-CMM tailored specifically to assessing and improving ESG capabilities. The proposed model draws on the theoretical principles of the original CMM while adapting them to the unique challenges and requirements of ESG.

The goal of this research is twofold: first, to provide organisations with a practical, structured tool to evaluate and enhance their ESG maturity; and second, to contribute to academic understanding in this relatively unexplored area. By combining theoretical rigor with real-world applicability, the ESG-CMM seeks to become a valuable resource for both practitioners and researchers.

1.1 Problem Statement and Research Question

Organisations across industries are under increasing pressure from stakeholders, including regulators, investors, and customers to adopt ESG practices. The participation is no longer optional for larger companies but has become essential for ensuring long term compliance with regulatory standards. However, despite the workload and importance of ESG, organisations do not have a systematic tool for evaluating ESG capabilities. This lack of structure could result in management having problems extracting the value of the department and limit the development and progress of the ESG practice in the organisation.

While CMMs have been successfully used in other fields, there is a noticeable gap in their application to ESG. Existing studies on CMM in relation to ESG areas are non-existent. This lack of research leaves organisations without a tool to assess their current ESG capabilities, identify gaps, and prioritise which areas to improve and put the most effort into. Thereby, the absence of an ESG-specific CMM presents a challenge for companies attempting to evaluate and enhance their ESG departments. Without defined maturity levels of the ESG capabilities, organisations rely on unstructured approaches. Addressing this gap in the existing research requires a development of a CMM specifically tailored for ESG that integrates insights from the academic knowledge of other areas and the information gathered from practical knowledge. Such a model could not just help to fill the research gap but also provide organisations with a tool for evaluating their ESG capabilities and drive the efforts for continuous improvement. We propose therefore the following research question and research objectives.

1.2 Research Question and Objectives

Research Question: *How can the theoretical framework capability maturity model be applied in an environmental social and governance context.*

The primary objective of this research is to develop a CMM tailored to ESG practices. This model has the objectives to provide organisations with a structured framework to evaluate the maturity of their ESG departments and identify areas for improvement to enhance their overall ESG capabilities.

Research objectives: (1) The development of an ESG-CMM; (2) Testing the ESG-CMM in real-world organisational contexts.

By meeting these objectives, the research aims to fill the gap in existing literature on CMM in relation to ESG and provide a tool that helps organisations improve their ESG practices effectively.

2 Research Method

This article adopts Design Science Research (DSR) [10] as its umbrella methodology to explicitly target the creation and evaluation of problem-solving artefacts in information systems contexts [10]. Table 1 provides the six steps of the DSR activities complimented by the overview of the process in Fig. 1.

At the core of the second and third step is the model development, building on existing literature and synthesising the findings into the ESG-CMM artifact.

A systematic literature search was conducted using Scopus and Web of Science using the string: ("capability maturity model" OR "maturity model") AND (sustainability OR governance OR risk OR "digital transformation"). This yielded 1.793 results. Further filtering based on limiting the output to articles in the "Engineering" and "Business Management and Accounting" fields as well as the years "2019–2025" reduced this to 82 articles. Screening (title–abstract then full text) and quality appraisal returned 15 high-quality articles spanning digital transformation, Safety, Health, and Environment (SHE) management, resilience, knowledge management and Information Communication Technology (ICT) literacy. From each article the parameters (n = 104) that authors used to operationalise maturity were extracted.

The 104 descriptors were subjected to open coding and comparison. Duplicate or synonymous terms were merged, producing 18 first-order themes. An axial-coding grouped

Table 1. Adopted Six-Activity Design Science Research (DSR) Process Based on [10].

DSR Activity	Application in the case
1 Problem identification & motivation	Gap analysis revealed no capability-maturity model (CMM) tailored to Environmental-Social-Governance (ESG) functions; companies therefore lack structured guidance for building ESG capability
2 Define objectives of a solution	Target outcomes: a transparent, theory-grounded ESG-CMM that is actionable for both engineering and management audiences. Foundations combine classic CMM, the Technology Organisation Environment (TOE) [11] lens and EU ESRS requirements
3 Design & development	Iterative creation of the ESG-CMM using abductive cycles of literature synthesis, semi-structured interviews and a single-case study MidManuCo (pseudonym)
4 Demonstration	Pilot application at MidManuCo: seven stakeholder interviews assessed model usability and generated an initial ESG maturity profile
5 Evaluation	Interview feedback and maturity scores were analysed to refine parameter definitions, clarify level descriptors and confirm practical fit
6 Communication	Interim model shared with interviewees; future practitioner rollout at MidManuCo

those themes by affinity into five transversal capability clusters (governance, technology, people, risk, improvement) [12]. To ensure coverage of contextual factors, the clusters were validated against the Technology Organisation Environment (TOE) framework [11]. This resulted in no additional themes being identified, confirming conceptual sufficiency. The final step condensed each cluster to one-to-three concise, objective ESG parameters (total = 13).

The parameters were arranged along the classic five-level CMM ladder: (1) Initial, (2) Repeatable, (3) Defined, (4) Managed and (5) Optimising, yielding the ESG-CMM artefact [6].

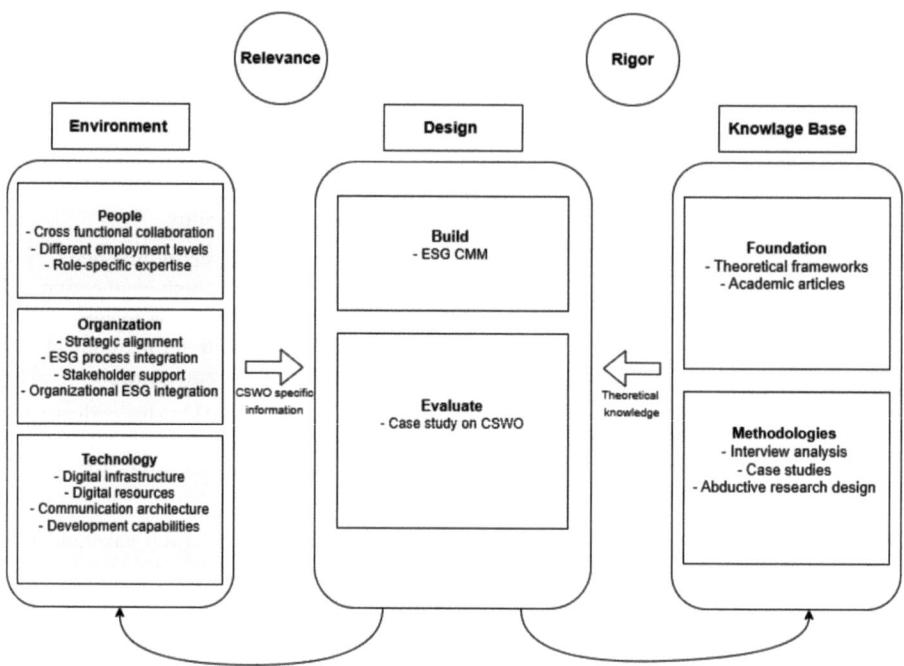

Fig. 1. Adapted DSR framework [10] for the development of the ESG-CMM artifact.

3 Related Work

A systematic literature review yielding 15 capability-maturity studies from sectors as diverse as digital transformation [13], safety–health–environment [14] and flood resilience [15] yielded 104 distinct capability descriptors. These were synthesised into 5 major clusters as described in the table below (Table 2).

Table 2. Clustering of themes based on the 15 selected articles

Cluster	Representative themes	Related work
Strategic governance & alignment	Policy & vision, leadership ownership, regulatory compliance	[13, 16, 17]
Technology & data infrastructure	IIoT/IT platforms, data quality, interoperability	[18–20]
People & culture	Workforce competence, collaboration, innovation mind-set	[21–23]
Risk & resilience management	Hazard identification, BCM, SHE, flood resilience	[14, 15, 24]
Process improvement & performance management	Continuous-improvement routines, KPI systems, SPI	[20, 25–27]

3.1 Strategic Governance and Alignment

This cluster bundles everything that gives direction, funding and legitimacy to improvement programmes. Typical descriptors are *Strategic Governance* and *Policy & Vision* [13], *Senior-Management Commitment to SHE* [14] and *Regulatory Compliance* [17]. A recurring insight is that governance mechanisms must translate external mandates (e.g., ESRS, ISO clauses) into an internal "why" and then allocate budgets and KPIs accordingly. In the ESG-CMM these ideas surface as Governance, Policy, Objectives and Budget Ownership parameters.

Parameters related to this cluster: Strategic Governance, Senior Management Commitment to SHE, SHE Policy, SHE Objectives & Targets, Definition of Responsibilities, Success Factors, Regulatory Compliance, IT Governance, Leadership, Vision & Goals, Business Model, Value-Realisation Planning, Governance, Framework, Equipment-Management Strategy, Management, Digital Leadership, Agile Process (governance layer), Resources.

3.2 Technology and Data Infrastructure

Here the literature concentrates on IT-OT capabilities that turn raw operations into decision-ready data. Examples include *Electronic Data Management* and *Data-flow Monitoring* [20], *IT Application Maturity* [16], *Equipment-Management IT Infrastructure* [19] and *Knowledge-Management Systems* [25]. Studies emphasise interoperability, real-time capture and analytics readiness. For the ESG-CMM we fold these ideas into IoT Data Infrastructure, Information Systems and Analytics Capability, parameters that later enable automated scoring of ESG metrics.

Parameters related to this cluster: Information & Technology, Electronic Data Management, Utility/Communication System, Technical Standardisation, IT Application Maturity, Data Flow Monitoring, Methods & Tools, Information & Analytics, Business Applications, Communication Technology, Equipment-Management IT Infrastructure, Knowledge Management (tech layer), ICT (infrastructure), Data Analysis & Utilisation, Data Management.

3.3 People and Culture

This is the largest cluster by count, reflecting the consensus that skills and mind-sets often make or break maturity trajectories. Key descriptors: *Workforce Management/Skills* [13], *Innovation Culture* [21] and [22], *Collaboration* and *Motivation* [18], plus digital-literacy measures in public-sector ICT studies [23]. Across domains, high-maturity cultures share three traits, (1) continuous learning, (2) cross-functional collaboration, and (3) psychological safety for experimentation. In the ESG-CMM these translate to Skills, Culture, Collaboration and Training parameters.

Parameters related to this cluster: Workforce Management, Culture, SHE Training, Employee Involvement in SHE, SHE Competence, Communications, Collaboration, Rationality & Truth, Motivation, HR Policies, Workplace Creativity, Literacy Level of Basic Computer Knowledge, Literacy Level of Computer Usage Skill, Literacy Level of Computer Utilisation, Literacy Level of Internet Basic Knowledge, Literacy Level of Internet Usage Skill, Literacy Level of Internet Utilisation, Training-Needs Analysis, Curriculum Development, Training Execution, Equipment-Management Personnel, Organisational Inhibitors, Training Inhibitors, Educational Inhibitors, Collaborative (mind-set), Collaboration Culture, Innovation Culture, Digital Mind-set, Digital Literacy, Value-Creation Skill.

3.4 Risk and Resilience Management

Descriptors centre on anticipating, absorbing and recovering from shocks, whether safety incidents, floods or supply-chain disruptions. *SHE Risk Management* and *Performance Monitoring* [14], *Awareness of Flood Risk* and *Post-Event Review* [15] and *Risk Analysis & Mitigation* [20] typify the set. Papers stress closed-loop learning: incidents feed back into risk registers, which feed new controls. ESG-CMM mirrors this with Risk Management, Compliance, Resilience and Incident Learning parameters.

Parameters related to this cluster: SHE Performance Monitoring, SHE System Auditing, Physical SHE Resources, Financial Resources for SHE, Awareness and Understanding of Flood Risk, Turn-over and Cash Flow Management, Post-event Review, Analysis and Management, Quality, Control over Risks, Continuous Improvement Initiatives, Response to Agility, Cross-Functional Collaboration, Risk Analysis and Mitigation, Execution, Value Realisation, Manufacturing Performance, Management Inhibitors, Technology Inhibitors, Progress, Risk Assessment.

3.5 Process Improvement and Performance Management

Finally, a line of work rooted in Software-Process Improvement [24] and CMMI studies foregrounds structured, metrics-driven change. Descriptors include *Process Improvement*, *Continuous-Improvement Initiatives* and *KPI Systems*. The unifying idea is that maturity climbs fastest when feedback cycles are explicit, data-based and owned by cross-functional teams. Accordingly, the ESG-CMM keeps Continuous Improvement and Performance Management as standalone parameters.

Parameters related to this cluster: Digital-Process Transformation, SHE Management Programme, SHE Risk Management, Management of Outsourced Services, Project-Management Capability, Process Innovation, Process Optimisation, Change Impact Analysis, Change Management Planning, Change Management Execution, Optimisation (general), Equipment-Management Organisation & Process, Process Improvement, Knowledge Management, Agile Process (execution layer), Digital Processes & Standardisation, Digital Transformation & Collaboration Tools.

4 ESG-CMM Development

4.1 Parameter Consolidation

The 104 descriptors from the literature were first normalised for synonymy, then grouped into the five transversal clusters.

Within each cluster, one-to-three descriptors were retained using three filters: (1) Frequency across the 15 source papers, (2) Traceability to ESRS disclosure topics, and

(3) Managerial relevance confirmed in interviews. The result is the following 13 objective parameters (Table 3).

Table 3. Shows how each of the 13 parameters is related to the respective cluster and then is further framed as a single, outcome-oriented capability statement.

Cluster	Parameter	Capability statement
Strategic governance & alignment	Governance & Policy	The board assigns clear ESG accountability and approves a formal policy aligned with ESRS
	Strategic Alignment	ESG targets are embedded in enterprise strategy and cascaded to functional OKRs
Technology & data infrastructure	Data Infrastructure	An IIoT-enabled data pipeline captures energy, waste and safety metrics in real time
	Information Systems	Integrated IT systems aggregate ESG, financial and operational data for reporting
	Analytics Capability	Advanced analytics generate predictive ESG insights that inform operational decisions
People & culture	Workforce Skills	Key staff possess certified competencies to implement and improve ESG processes
	Culture & Collaboration	Cross-functional teams routinely collaborate on ESG initiatives and share lessons
	Training & KM	A structured programme builds ESG knowledge and retains know-how in a KM platform.
Risk & resilience management	Risk Management	Material ESG risks are identified, scored and mitigated through a live risk register
	Regulatory Compliance	Processes ensure timely, accurate submission of all mandatory ESG disclosures
	Resilience & Business Continuity	BCP scenarios include ESG disruptions and are validated through drills
Process improvement & performance management	Continuous Improvement	PDCA cycles drive incremental ESG performance gains organisation wide

(continued)

Table 3. (*continued*)

Cluster	Parameter	Capability statement
	Performance Management	*Leading and lagging ESG KPIs are monitored in near-real time and reviewed monthly*

4.2 Level Structure and Descriptors

Further, the Environmental-Social-Governance Capability-Maturity Model (ESG-CMM) embeds the 13 parameters into the classic five-level ladder: (1) Initial, (2) Repeatable, (3) Defined, (4) Managed and (5) Optimising, so that organisations can quickly locate their present capability and see the next logical step forward.

The five maturity levels reuse familiar CMM semantics but are tailored to ESG, the list of adapted maturity levels is below:

- **Level 1 – Initial**: Ad-hoc, undocumented ESG activities; data live in spreadsheets; success depends on individual effort.
- **Level 2 – Repeatable**: Basic policies exist, and routine data capture is scheduled, yet processes differ by department and metrics are lagging.
- **Level 3 – Defined**: Cross-functional procedures are documented, IoT gateways begin feeding energy- and emissions-sensor data into a central repository, and training is formalised.
- **Level 4 – Managed**: Quantitative targets, leading KPIs and automated alerts are in place; analytics dashboards combine financial and ESG performance for decision support.
- **Level 5 – Optimising**: Predictive analytics guide proactive action, real-time sensor data validate improvements, and lessons learned loop into quarterly strategy reviews.

4.3 The ESG-CMM Artifact

The finished ESG-CMM artifact is presented in the table below, broken into the 13 parameters and the 5 maturity levels (Table 4).

Table 4. ESG-CMM artifact

Parameter	Maturity 1 Initial	Maturity 2 Repeatable	Maturity 3 Defined	Maturity 4 Managed	Maturity 5 Optimised
Data analysis and utilisation	Analysis unstructured, ad-hoc, experience-based; collection inconsistent, rarely actionable	Foundations exist; decisions use simple metrics yet lack depth and insight	Analysis standardised and embedded; teams use data to align decisions with objectives	Advanced tools track KPIs; decisions are predictive and fully data-driven	AI/ML gives proactive insights; every decision is data-driven, fuelling innovation

(*continued*)

Table 4. (*continued*)

Parameter	Maturity 1 Initial	Maturity 2 Repeatable	Maturity 3 Defined	Maturity 4 Managed	Maturity 5 Optimised
Data management	Storage ad-hoc, poorly documented; access inconsistent, error-prone	Basic management: defined storage, some standardisation, better access	Standardised, enterprise-wide management with consistent security, storage and retrieval	Real-time tools monitor quality, security and performance	Governance improves via predictive analytics and feedback loops
Digital Processes and standardisation	Manual, disorganised, undocumented processes; errors frequent	Some documentation, basic digital tools; standardisation limited	Processes documented and standardised, backed by digital tools	Integrated platforms automate and optimise for efficiency	AI/ML-driven automation enables adaptive, seamless workflows
Digital Transformation and collaboration tools	Digital tools rare; collaboration manual, siloed; remaining tools outdated or unused	Basic tools allow structured communication, but use inconsistent	Tools standardised across teams, embedded in workflows	Real-time, cross-functional collaboration integrates with other systems	AI-driven platforms foster proactive, innovative teamwork
ESG influence in company decisions	ESG absent or reactive; decisions chase short-term goals, ignore impacts	ESG considered in some decisions but lacks integration and impact	ESG formally integrated, aligned with policies, goals and strategy	ESG guides decisions; KPIs measured against clear targets	ESG drives innovation; every decision follows long-term strategy
Management	No clear ESG direction; department isolated, minimal guidance.	ESG recognised yet support limited; department reactive, weakly aligned	Structured processes align ESG with goals; priorities and accountability clear	ESG embedded in operations, consistently supported and resourced	ESG fully embedded, enabling seamless collaboration and evolving goals
Corporate culture	ESG awareness low; engagement obligatory; department dismissed	Visibility grows but awareness limited; department a "necessary evil."	Department seen as key resource for compliance and ESG	Department integral; staff rely on its expertise for ESG targets	Department leads strategy; employees proactively collaborate on ESG
Resources	Minimal resources limit department	Basic support arrives but capacity remains constrained	Adequate backing enables effective, aligned work	Consistent resources deliver impactful results	Strategic backing empowers innovation and leading initiatives

(*continued*)

Table 4. (*continued*)

Parameter	Maturity 1 Initial	Maturity 2 Repeatable	Maturity 3 Defined	Maturity 4 Managed	Maturity 5 Optimised
Governance	ESG governance absent; no oversight	Basic governance meets minimum compliance; oversight inconsistent	Formal governance ensures compliance and goal alignment	Integrated governance monitors and enforces accountability	Proactive governance drives improvement, aligned with global standards
Progress	Minimal, reactive progress; performance unmeasured	Sporadic progress; basic tracking, limited innovation	Structured progress aligned with goals; performance measured	Systematic progress; metrics guide continuous advancement	Progress drives excellence via benchmarking and predictive analysis
Impact, risk and opportunity management	No structured ESG risk approach; actions reactive	Basic processes handle immediate risks; opportunities rarely explored	Systematic risk management aligns with goals; proactive mitigation	Integrated tools give real-time assessments and strategic leverage	Predictive models and input drive innovation and resilience
Internal stakeholder engagement & comm.	Engagement limited, reactive; communication uncoordinated	Basic engagement processes inconsistent; collaboration sporadic	Structured engagement; stakeholders receive timely, goal-aligned information	Proactive collaboration embeds engagement in workflows	Optimised engagement builds trust, sparks innovation, aligns stakeholders
Assurance and Regulatory Alignment	Alignment patchy; no formal assurance, risking non-compliance	Basic measures under pressure; assurance inconsistent	Integrated alignment with structured assurance; outcomes reliable	Embedded compliance monitored regularly, ensuring improvement	Predictive tools exceed standards, driving excellence

5 Case Study: MidManuCo

5.1 Organisational Context

MidManuCo (pseudonym) is a privately-owned, heavy-manufacturing firm with roughly 800–1000 employees spread across production, engineering and supporting functions. The company fabricates large steel components for the energy-infrastructure market and, like many European organisations, will fall under the expanded ESRS disclosure regime from 2028.

5.2 Data-Collection

Seven semi-structured interviews were conducted with domain leads from Compliance & ESG (1 & 2), Production, IT, HR, Finance and Quality. The interview guide comprised three parts: (1) Parameter rating, respondents scored each of the 13 ESG-CMM parameters on the five-level rubric using anchor descriptors drawn from Table 1. (2) Evidence probe, for every score, interviewees were asked to cite tangible artefacts (policies, dashboards, sensor logs, audit reports). (3) Barrier/enabler narrative, open questions captured success factors, pain points and improvement ideas.

5.3 Maturity Results

The average across all parameters is 2.0, placing MidManuCo at the Repeatable level. Dispersion is moderate ($\sigma = 0.46$), implying pockets of good practice already exist. The cluster view is provided in the list in the table below (Table 5).

Table 5. The table shows the scoring of the case with some example statements.

Cluster	Score	Example statement
Governance & Alignment	2.4	A formal ESG policy has board endorsement and Scope-2 emission targets are tracked monthly, yet roles remain implicit
Technology & Data	1.8	The data analyst captures machine-energy data at five-second intervals, but workflows still rely on manual exports and spreadsheet macros; analytics is descriptive rather than predictive.
People & Culture	2.2	Front-line and support staff have received basic climate-literacy training. Cross-functional collaboration occurs only when customers demand audits
Risk & Resilience	1.7	Legal-compliance tasks are executed, but an integrated ESG risk register and BCM scenarios are missing
Improvement & Performance	2.0	The continuous-improvement programme addresses scrap and throughput but rarely links improvements to ESG KPIs
Mean	2.0	

Following the data collection, open coding produced 28 barrier/enabler fragments, merged into six themes (Table 6).

Table 6. The table shows the synthesis of 28 barriers/enablers merged into 6 themes.

Theme	Description	Frequency
Fragmented data landscape	Sensor feeds in historian; financial data in ERP; ESG metrics in Excel, no "single source of truth"	6
Siloed responsibilities	Compliance "owns" reporting, but improvement actions sit with production; weak feedback loops	5

(*continued*)

Table 6. (*continued*)

Theme	Description	Frequency
Limited analytics know-how	Dashboards exist, yet staff rely on static charts; no data scientist on payroll	4
Executive-time scarcity	ESG competes with production KPIs for C-suite attention; funding cycles are quarterly	4
Supplier opacity	>70% of steel mass is external; suppliers provide only annual footprint updates	3
Positive customer pull	Key buyers willing to co-fund digital-meter pilots if data quality improves	3 (enabler)

Figure 2 provides a visual heatmap of the interview responses, illustrating how each of the 13 ESG-CMM parameters scored across the five clusters. Darker shades represent higher maturity levels, helping to pinpoint capability concentrations and gaps. This visual reinforces the scoring summary, making it easier to identify underdeveloped areas such as analytics and integrated risk management, and to prioritise actions for improvement.

Bridging ESG and Capability Maturity 201

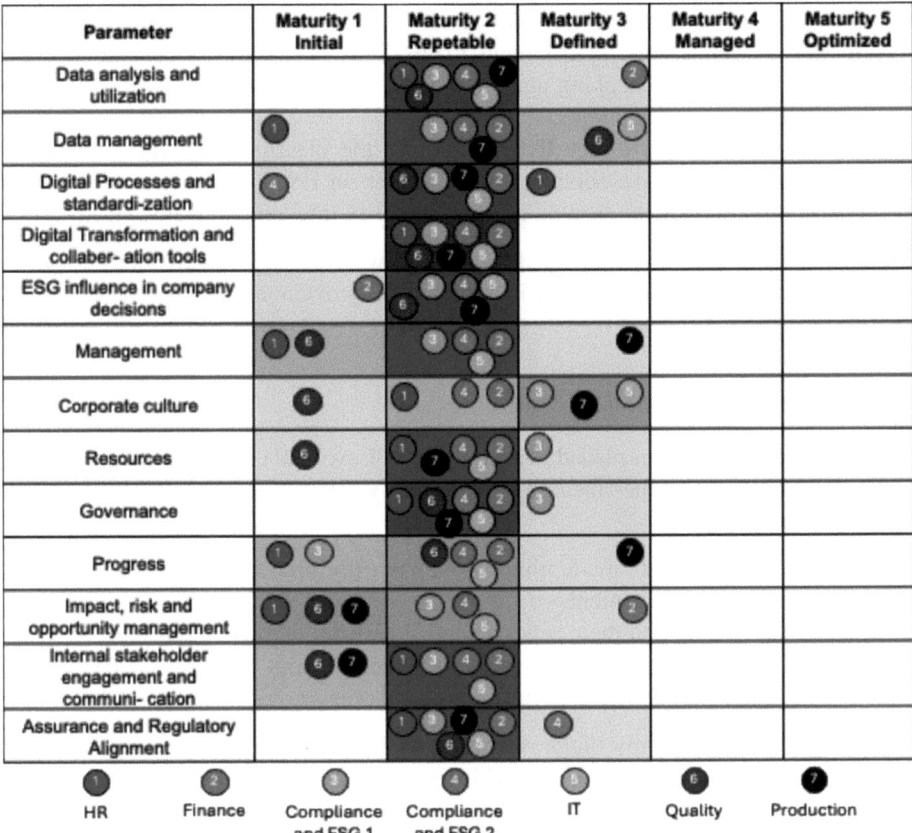

Fig. 2. Heatmap of interview responses coded into the ESG-CMM artefact.

6 Discussion

This article set out to explore how a theoretical framework called CMM could be applied in an ESG context. By adapting the foundational principles of the CMM, traditionally used to in the software area, the research developed a structured model tailored to ESG practices. The resulting ESG-CMM model provided a systematic framework to assess ESG capabilities, identify areas for improvement, and foster continuous development.

6.1 Adapting CMM to ESG

During the adaptation process, three key factors provided the main logic behind the model development:

1. **Parameter derivation from literature.** A systematic review of 15 maturity-model studies supplied 104 raw descriptors. Grounded-theory coding merged these into 18 themes and five transversal capability clusters.

2. **Contextualisation with TOE and ESRS.** Inspecting each theme with the Technology-Organisation-Environment (TOE) lens [11] ensured contextual completeness, while cross-checking against the EU's ESRS disclosure topics anchored the artefact in current regulation.
3. **Level wording and evidence lists.** Retaining the classic five CMM levels preserved familiarity, but descriptors were rewritten in ESG. Each parameter carries 2–3 observable indicators, turning abstract maturity into auditable evidence.

Together these developments transformed a software-centric framework into a structured ESG roadmap without losing the incremental-improvement DNA that makes CMM effective.

6.2 Empirical Confirmation at MidManuCo

Interview-based assessment placed MidManuCo at Level 2.0 (Repeatable), confirming that the artefact distinguishes meaningful capability stages even in a single-site pilot. Respondents reported that:

- The ESG-CMM artifact "made blind spots painfully obvious", particularly around analytics and risk management.
- The evidence lists helped them separate anecdote from fact when scoring.
- The ladder metaphor triggered constructive debate about *sequencing*, e.g., fixing governance gaps before heavy technology spends.

Thus, the model not only diagnoses maturity but also stimulates cross-functional dialogue, one of the original aims of CMM in software.

6.3 Implications for Research and Practice

Research Contribution. The ESG-CMM fills a documented gap: no peer-reviewed maturity model previously linked ESG requirements to CMM theory. Our work shows that legacy frameworks can be re-engineered for new problem spaces through systematic parameter synthesis and contextual lenses.

Practical Contribution. For businesses bracing for ESRS, the model offers a self-assessment tool that is (i) regulator-aligned, (ii) expressed in plain business language, and (iii) scalable, from spreadsheet check-ups to full IoT-enabled dashboards.

6.4 Limitations and Future Work

Although the single-case design provides rich insight, it inevitably limits the model's generalisability, and the interview-based ratings, despite being supported by evidence, still carry an element of subjectivity. Future studies should therefore replicate the ESG-CMM pilot across multiple industries to test its robustness.

7 Conclusion and Future Work

This study demonstrates that the Capability Maturity Model (CMM), traditionally applied in software engineering, can be successfully adapted to assess and guide Environmental, Social, and Governance (ESG) practices in industrial organisations. The resulting ESG-CMM offers a structured, regulator-aligned framework that allows companies to identify their current ESG maturity, uncover capability gaps, and prioritise improvement efforts. The single-case application at MidManuCo confirmed the model's usability and diagnostic value, placing the organisation at Level 2 (Repeatable) and exposing critical areas for development, especially in analytics and integrated risk management. A notable insight was how the model's maturity levels and evidence-based scoring encouraged cross-functional dialogue, making ESG more actionable and strategically embedded.

A key lesson learned is that re-engineering legacy frameworks like CMM for new domains requires balancing theoretical foundations with practical relevance. Anchoring the model in ESRS requirements and organisational realities ensured it was not only academically rigorous but also implementable. Furthermore, the pilot revealed that even basic self-assessment tools can serve as catalysts for internal reflection, organisational alignment, and ESG momentum. Looking ahead, a significant opportunity lies in linking the ESG-CMM to Industrial Internet of Things (IIoT) technologies. Real-time data capture from sensors can enhance the accuracy and responsiveness of ESG tracking, enabling predictive insights and automation of performance measurement. Future research should focus on multi-site validation of the model and explore extensions into sensor-driven, real-time ESG maturity scoring to further operationalise sustainable transformation.

References

1. Jámbor, A., Zanócz, A.: The diversity of environmental, social, and governance aspects in sustainability: a systematic literature review. Sustainability **15**(18), 13958 (2023)
2. Pantazi, T.: The introduction of mandatory corporate sustainability reporting in the EU and the question of enforcement. Eur. Bus. Organ. Law Rev., 1–24 (2024)
3. European Parliament & Council. Directive (EU) 2022/2464 of 14 December 2022 as regards corporate sustainability reporting (CSRD). Official Journal L 322, pp. 15–80, 16 December 2022
4. Humphrey, W.S.: Managing the Software Process. ADDISON-Wesley Longman Publishing Co., Inc. (1989)
5. Paulk, M.C., Curtis, B., Chrissis, M.B., Weber, C.V.: Capability maturity model for software. Carnegie Mellon University, Software Engineering Institute, Pittsburgh, PA, USA (1991)
6. Paulk, M.C., Curtis, B., Chrissis, M.B., Weber, C.V.: Capability maturity model, version 1.1. IEEE Softw. **10**(4), 18–27 (1993)
7. Technical Presentation: CMMI for Systems/Software Engineering, version 1.0, CMU (2000)
8. Technical Presentation: CMMI for Software Engineering, version 1.1, CMU (2002)
9. Technical Report: CMMI® for Development, Version 1.3, CMU (2006)
10. Peffers, K., Tuunanen, T., Rothenberger, M.A., Chatterjee, S.: A design science research methodology for information systems research. J. Manag. Inf. Syst. **24**(3), 45–77 (2007)
11. Baker, J.: The technology–organization–environment framework. Inf. Syst. Theory Explain. Predict. Our Digit. Soc. **1**, 231–245 (2011)

12. Corbin, J., Strauss, A.: Basics of Qualitative Research: Techniques and Procedures for Developing Grounded Theory. Sage Publications, Thousand Oaks (2014)
13. Gökalp, E., Martinez, V.: Digital transformation maturity assessment: development of the digital transformation capability maturity model. Int. J. Prod. Res. **60**(20), 6282–6302 (2022)
14. Asah-Kissiedu, M., Manu, P., Booth, C.A., Mahamadu, A.M., Agyekum, K.: An integrated safety, health and environmental management capability maturity model for construction organisations: a case study in Ghana. Buildings **11**(12), 645 (2021)
15. Adeniyi, O., Perera, S., Ginige, K., Feng, Y.: Developing maturity levels for flood resilience of businesses using built environment flood resilience capability areas. Sustain. Cities Soc. **51**, 101778 (2019)
16. Liao, R., Chen, H., Sun, C., Sun, Y.: An exploratory study on two-dimensional project management maturity model. Eng. Manag. J. **35**(4), 445–459 (2023)
17. Jain, A., Bruckmann, D., van der Heijden, R.E., Marchau, V.A.: Towards rail-related multimodal freight exchange platforms: exploring regulatory topics at EU level. Compet. Regul. Netw. Ind. **20**(2), 138–163 (2019)
18. Tsai, W.L.: The impact of project teams on CMMI implementations: a case study from an organizational culture perspective. Syst. Pract. Action Res. **34**, 169–185 (2021)
19. Han, S., Li, C., Feng, W., Luo, Z., Gupta, S.: The effect of equipment management capability maturity on manufacturing performance. Prod. Plan. Control **32**(16), 1352–1367 (2021)
20. Thomas, T., Saleeshya, P.G.: CMMI based fuzzy logic approach to assess the digital manufacturing maturity level of manufacturing industries. TQM J. **35**(8), 2658–2683 (2023)
21. Lee, J.C., Chen, C.Y.: The moderator of innovation culture and the mediator of realized absorptive capacity in enhancing organizations' absorptive capacity for SPI success. J. Glob. Inf. Manag. (JGIM) **27**(4), 70–90 (2019)
22. Devi, E.T., Wibisono, D., Mulyono, N.B.: Identifying critical capabilities for improving the maturity level of digital services creation process. J. Ind. Eng. Manag. **15**(3), 498–519 (2022)
23. Waluyo, R., Gunawan, F.E., Setiawan, I.: The measurement of information and communication technology literacy: a case study of the village officials in Purbalingga. CommIT (Commun. Inf. Technol.) J. **16**(1), 19–25 (2022)
24. Putri, N.K.S., Permatasari, D., Susanto, R., Lee, C.K., Kurniawan, Y.: Knowledge management evaluation using digital capability maturity model in higher education institution. Electron. J. Knowl. Manag. **21**(2), 140–157 (2023)
25. Shaikh, A.: The role of maturity driven software process improvement in an industry. Int. J. Adv. Trends Comput. Sci. Eng. **8**(11), 344–350 (2019)
26. Giménez-Medina, M., Enríquez, J.G., Olivero, M.A., Domínguez-Mayo, F.J.: The innovation challenge in Spain: a Delphi study. Expert Syst. Appl. **230**, 120611 (2023)
27. Harin, R.S., Sanmukhaprian, S.R.J., Raghuram, P., Sreedharan, V.R., Zouadi, T.: Developing a capability maturity model for knowledge exchange dynamics among stakeholders and universities: evidence from Morocco higher education institution. IEEE Trans. Eng. Manag. (2023)

Open Access This chapter is licensed under the terms of the Creative Commons Attribution 4.0 International License (http://creativecommons.org/licenses/by/4.0/), which permits use, sharing, adaptation, distribution and reproduction in any medium or format, as long as you give appropriate credit to the original author(s) and the source, provide a link to the Creative Commons license and indicate if changes were made.

The images or other third party material in this chapter are included in the chapter's Creative Commons license, unless indicated otherwise in a credit line to the material. If material is not included in the chapter's Creative Commons license and your intended use is not permitted by statutory regulation or exceeds the permitted use, you will need to obtain permission directly from the copyright holder.

An NGSI-LD-Based ICT Tool for Data Visualization and Traceability in Sustainable Supply Chains and Biological Resource Certification

Romain Magnani(✉) and Franck Le Gall

EGM, 06560 Valbonne, France
{romain.magnani,franck.le-gall}@egm.io

Abstract. This paper presents a novel approach to biological resource traceability and certification through an IoT-ready platform based on NGSI-LD data models. The BioReCer ICT Tool (BIT) addresses significant challenges in sustainable supply chain management by providing an interoperable, standardized framework for data collection, visualization, and verification. Our implementation leverages the Stellio Context Broker together with Apache NiFi for Internet of Things (IoT) data integration, creating a powerful platform for sustainability assessment and certification. Both the data model and the user interface have been co-developed and validated with stakeholders across four distinct case studies, demonstrating its flexibility and effectiveness across diverse bio-resource value chains. Results indicate significant improvements in data interoperability, traceability, and verification capabilities compared to traditional approaches.

Keywords: Supply chain · bio-resources · NGSI-LD · Context Broker · sustainability · IoT · interoperability · data visualization · user experience

1 Introduction

1.1 Purpose of This Paper

The digital transformation of supply chains has created new opportunities for improving sustainability assessment and certification processes, particularly for biological resources. However, current approaches suffer from data heterogeneity, interoperability challenges, and lack of standardization. This paper introduces an innovative tool for Supply Chain and Sustainable Biological Resources Certification data visualization and traceability based on a co-validated NGSI-LD data model. The NGSI-LD information model is a property graph model, particularly convenient for process tracking. The proposed solution enables comprehensive data collection, analysis, and verification throughout complex bio-resource value chains, supporting more transparent sustainability assessment and certification processes.

The adoption of NGSI-LD as the underlying data model specification provides a standardized approach to context information management and facilitates seamless integration of heterogeneous data sources. This approach addresses critical challenges in current certification systems by enabling interoperability between different IoT platforms, sensors, and data sources, while providing a unified interface for data visualization and traceability.

1.2 BioReCer Project Overview

The BioReCer (Biological Resources Certification) project aims to develop and validate innovative tools for assessing the sustainability of bio-resources throughout their value chains, with particular emphasis on developing appropriate indicators to measure and assess their circularity [1]. The project implements a quintuple helix innovation model that engages stakeholders from academia, industry, government, civil society, and the environment to co-create effective solutions for sustainability assessment and certification.

At the core of the BioReCer project, the BioReCer ICT Tool (BIT) functions as a web portal for data collection, analysis, and visualization related to sustainability assessment and certification. By focusing on the entire value chain, BioReCer acknowledges the interconnected nature of bio-based product lifecycles, where the feedstock for one actor is the product of another. This approach is crucial for maintaining circularity throughout the product lifecycle and forms the foundation for the technological solutions developed within the project, including the BIT platform.

The project utilizes four case studies as testbeds to validate the effectiveness of the proposed methodologies and tools. These case studies span diverse bio-resource value chains, ensuring the applicability and adaptability of the developed solutions across different contexts and sectors.

1.3 Project Technical Challenges

The development of the BIT platform presents several significant technical challenges related to data management, interoperability, and visualization across complex biological resource supply chains. First, the heterogeneous nature of data sources in bio-based supply chains—ranging from farm management systems to manufacturing processes and waste management facilities—requires robust data integration capabilities that can accommodate various formats, protocols, and temporal resolutions while maintaining data integrity and context.

Second, the project faces the challenge of creating a standardized yet flexible data model that can represent the diverse characteristics of bio-based products and their circularity metrics. This model must capture not only the physical and chemical properties of materials but also their biological attributes, lifecycle stages, and sustainability impacts, requiring an advanced semantic approach to data representation.

Third, ensuring interoperability between existing certification schemes, company information systems, and the BioReCer platform demands careful consideration of technical standards and communication protocols. The system must be capable of exchanging data with external databases, IoT devices, and certification bodies while preserving the semantic meaning of the information throughout these interactions.

Fourth, visualizing complex relationships between supply chain actors, material flows, and circularity indicators in an intuitive and meaningful way presents significant user interface challenges. Visualization must balance technical accuracy with usability for diverse stakeholders, from technical experts to non-specialist users, requiring thoughtful design and user experience considerations.

Finally, the implementation of secure data sharing mechanisms that protect sensitive business information while promoting transparency for certification purposes requires sophisticated access control and data privacy solutions. The platform must navigate the tension between open data principles and proprietary information protection, particularly in competitive market environments.

2 IoT Applications in Supply Chain Management

2.1 A Review of Existing Tools

The integration of Internet of Things (IoT) technologies in supply chain management has evolved significantly in recent years, with numerous tools and platforms emerging to address visibility, traceability, and sustainability challenges. Current state-of-the-art solutions can be categorized into several overlapping domains relevant to bio-based supply chains.

RFID-based tracking systems represent one of the most established approaches, with applications such as IBM's Food Trust [2] and Walmart's Food Traceability Initiative [3] demonstrating the feasibility of real-time product tracking throughout supply chains. These systems typically combine RFID tags, readers, and middleware with cloud-based data management platforms to provide visibility into product movements and transformations. While effective for physical tracking, these solutions often lack the semantic richness required for comprehensive sustainability assessment of bio-based products.

Blockchain-enabled supply chain platforms, including Provenance [4], VeChain [5], and Origin Trail [6], have gained prominence for their ability to create immutable records of transactions and transformations across supply networks. Tian [7] demonstrated that blockchain technology, when combined with IoT data collection, can significantly enhance food supply chain transparency and build consumer trust. However, as noted by Kouhizadeh et al. [8], these systems typically focus on transaction verification rather than comprehensive environmental impact assessment, limiting their standalone value for circularity certification.

Environmental monitoring IoT networks, exemplified by systems like Libelium's Smart Agriculture solution [9] and Bosch's Environmental Monitoring System [10], provide continuous data collection on parameters relevant to sustainable agriculture and processing, including soil conditions, water usage, energy consumption, and emissions. These tools generate valuable data for sustainability assessment but often exist in isolation from broader supply chain management systems, creating integration challenges.

Digital Twin platforms, such as Siemens' MindSphere [11] and Microsoft's Azure Digital Twins [12], offer virtual representations of physical supply chains, enabling simulation, optimization, and real-time monitoring. Boje et al. [13] highlights the potential of Digital Twins to model complex biological systems and their interactions with technical processes, making them particularly relevant for bio-based supply chains. However, these platforms typically require substantial customization to incorporate specialized circularity metrics for bio-based products.

Notably, few existing tools effectively combine IoT capabilities with comprehensive circularity assessment specifically designed for bio-based products. The European Bio-based Industries Consortium [14] has identified this gap as a significant barrier to market adoption of sustainable bio-based alternatives, highlighting the need for specialized technological solutions like those proposed in the BioReCer project.

2.2 Technical and Social Challenges

The implementation of IoT solutions for sustainable supply chain management in the bio-based sector faces distinctive technical and social challenges that must be addressed to achieve widespread adoption and impact.

From a technical perspective, data interoperability remains a critical challenge due to the diverse standards, protocols, and data formats used across agricultural, processing, and retail systems. As noted by Wolfert et al. [15], the fragmentation of data landscapes in agricultural and bio-based supply chains creates significant integration hurdles. This is particularly problematic for circularity assessment, which requires continuous data flow across organizational boundaries. Current solutions often struggle with semantic interoperability, where the meaning and context of data are preserved when shared between different systems.

Connectivity limitations present another significant barrier, especially in primary production settings. Rural agricultural areas frequently lack reliable broadband infrastructure, creating data transmission bottlenecks that compromise real-time monitoring capabilities. This challenge is exacerbated by the diverse geographical distribution typical of bio-based supply chains, which often span multiple regions with varying levels of technological infrastructure.

Data quality and consistency issues arise from the inherent variability of biological systems. Unlike technical supply chains with standardized components and processes, bio-based supply chains must account for natural variations in feedstock properties, seasonal fluctuations, and biological degradation processes. Existing IoT systems typically lack the sophisticated data validation and normalization capabilities required to account for these biological variabilities.

From a social perspective, privacy concerns and competitive sensitivities present significant adoption barriers as shown by the first results of the BioReCer project, during user testing sessions and focus group with stakeholders. Bio-based supply chains often involve proprietary processes, formulations, and efficiency metrics that companies are reluctant to share. El Bilali and Allahyari [16] identify information sharing reluctance as a major barrier to transparency initiatives in food and agricultural supply chains. This challenge is particularly acute for sustainability certification, which requires detailed process data that may reveal competitive advantages or disadvantages.

Digital literacy disparities across supply chain participants create additional implementation challenges. While large industrial processors may have sophisticated data management capabilities, smaller producers and waste management operators often lack the technical expertise and resources to implement and maintain IoT systems. This creates digitalization bottlenecks at critical points in the circular value chain, compromising the completeness and reliability of lifecycle data.

Finally, the cost-benefit perception of IoT implementation represents a significant adoption barrier. Many stakeholders in bio-based supply chains operate with thin margins and limited investment capacity. Without clear demonstration of economic returns from improved certification processes, these actors may be reluctant to invest in sophisticated monitoring and data sharing capabilities.

3 NGSI-LD for Context-Rich Data Representation

3.1 The NGSI-LD Specification

NGSI-LD [17] represents a significant advancement in the field of context information management, providing a standardized API and data model specifically designed to facilitate the representation, access, and exchange of contextual information in heterogeneous environments. Developed by ETSI (European Telecommunications Standards Institute), NGSI-LD extends the earlier NGSI specification by incorporating linked data principles, allowing for richer semantic representation of entities and their relationships.

The core innovation of NGSI-LD lies in its incorporation of JSON-LD (JavaScript Object Notation for Linked Data) as its underlying data format, combining the developer-friendly aspects of JSON with the semantic richness of RDF (Resource Description Framework). This approach enables the creation of "knowledge graphs" that explicitly represent not just entities and their attributes, but also the relationships between them, along with temporal and spatial context. This capability is particularly valuable for bio-based supply chains, where understanding the relationships between feedstocks, processes, products, and environmental impacts is crucial for meaningful circularity assessment.

NGSI-LD's data model revolves around several key concepts that enable its context-rich representation. Entities represent the core information objects (such as feedstocks, processes, or products in a bio-based supply chain), each with a unique identifier. Properties describe the characteristics of entities (such as weight, composition, or sustainability metrics), while Relationships explicitly connect entities to each other (such as "is derived from" or "is processed by"), creating a navigable graph of interconnected information. Importantly, NGSI-LD supports Property-of-Property constructs, allowing attributes to have their own sub-attributes, such as provenance information, uncertainty measures, or timestamps—critical for scientific data management in circularity assessment.

The specification also provides sophisticated temporal capabilities, enabling the representation of time series data, temporal relationships, and historical context information. This feature is particularly relevant for life cycle assessment and circularity metrics, which must account for changes in material properties and environmental impacts over time. Similarly, NGSI-LD's geospatial capabilities allow for the representation of location information using standardized formats, facilitating spatial analysis of material flows and environmental impacts.

The query language capabilities of NGSI-LD represent another significant advantage for complex data retrieval scenarios. The specification supports filtering based on entity types, attribute values, geographical scope, and temporal parameters, enabling precise queries that can extract contextually relevant information from large datasets. This capability is essential for certification processes that must evaluate specific circularity criteria across complex supply chains.

Finally, NGSI-LD's subscription and notification mechanisms enable event-driven architectures where systems can receive updates when context information changes, facilitating real-time monitoring and responsive decision-making in dynamic supply chain environments. Within the framework of the BioReCer project, this feature is leveraged to keep the product's Circularity Indicators calculations updated in real-time.

3.2 Stellio Context Broker: A Powerful API for Tackling Identified Technical Challenges

The Stellio Context Broker [18] represents an implementation of the NGSI-LD specification, offering enhanced capabilities that address several of the technical challenges previously identified. As an open-source component developed within the FIWARE ecosystem, Stellio provides a robust API for managing context information with particular strengths in handling complex, interconnected data models—a critical requirement for representing bio-based supply chains and their circularity metrics.

Stellio's architectural design is optimized for graph-based data representation, utilizing PostgreSQL as its underlying storage technology. This design choice enables efficient traversal of relationship chains in complex supply networks, allowing the system to quickly answer queries that require navigating multiple links between entities—such as tracing a finished product back to its original biological feedstock while accounting for all transformation processes and inputs along the way. This capability directly addresses the challenge of maintaining material traceability throughout complex bio-based value chains.

The context broker implements comprehensive NGSI-LD temporal query capabilities, supporting both property-by-property time series and versioning of complete entities. This temporal awareness enables the representation of dynamic aspects of bio-based supply chains, such as seasonal variations in feedstock properties, batch-specific processing parameters, and evolving product characteristics. For circularity assessment, this temporal dimension is essential for accurately calculating time-dependent metrics such as degradation rates, shelf life, and resource efficiency over time.

Stellio's batch operation capabilities provide significant performance advantages when dealing with the high-volume data typical of IoT-enabled supply chains. The broker efficiently handles batch entity creation, update, and query operations, reducing network overhead and processing time when managing large collections of context data. This efficiency is particularly valuable when integrating data from numerous IoT sensors across distributed supply chain operations.

The context broker also implements advanced access control mechanisms that support fine-grained permissions at different scope levels such as the use case level, the entity type level, and the entity level. This capability addresses the challenge of balancing transparency for certification purposes with the protection of sensitive business information, allowing supply chain actors to selectively share specific sustainability metrics without exposing proprietary process details or commercial relationships.

Stellio's implementation of the NGSI-LD federation capabilities enables the creation of distributed context information systems where multiple brokers can exchange data according to defined governance rules. This federation approach supports the decentralized nature of bio-based supply chains, where different actors may maintain their own information systems while still participating in broader certification frameworks.

From an interoperability perspective, Stellio provides adapters for various data formats and protocols, facilitating integration with existing supply chain management systems, IoT platforms, and certification databases. The broker supports standard authentication methods (including OAuth2) and offers a comprehensive REST API that follows OpenAPI specifications, simplifying integration with third-party applications and services.

4 The BIT Architecture and Implementation

4.1 The NGSI-LD Context Data Model for Supply Chain

The NGSI-LD context data model developed for the BioReCer project represents a comprehensive ontological framework specifically designed to capture the complex relationships and attributes relevant to bio-based supply chains and their circularity assessment. Designed to be as generic as possible to answer the needs of different kinds of use cases, it serves as the architectural backbone of the BIT, organizing critical data into a structured knowledge graph comprised of interconnected entities, attributes, and relationships. The model has undergone significant refinement along the years to better align with evolving requirements and new results, introducing new entities and relationships to enhance functionality.

At the foundation of the model lies a set of core entity types that represent the physical components of bio-based supply chains. The Fig. 1 shows the NGSI-LD context data model where the blue rectangles represent entities (or Digital Twins), and white diamond-shapes represent relationships between them. Specific entities' properties have been removed for readability, the full model is too large to be presented in detail within this document; however, it is important to visualize the overall structure and how each entity connects with each other.

To enhance readability, the model is cut into smaller sectors. The following tables are describing entity types by sector, A, B and C (Table 1).

Considering the technical challenges of data gathering with connectivity limitations, digital literacy disparities, consistency issues, limited investment capacity, disparities across supply chain participants, etc., complete information on every transporter and individual travel instance is unlikely to be available. To address this, we've decided to see transporter travels as archetypes, requiring only data on recurring travel patterns within a company. This approach assumes that companies optimize their logistics, resulting

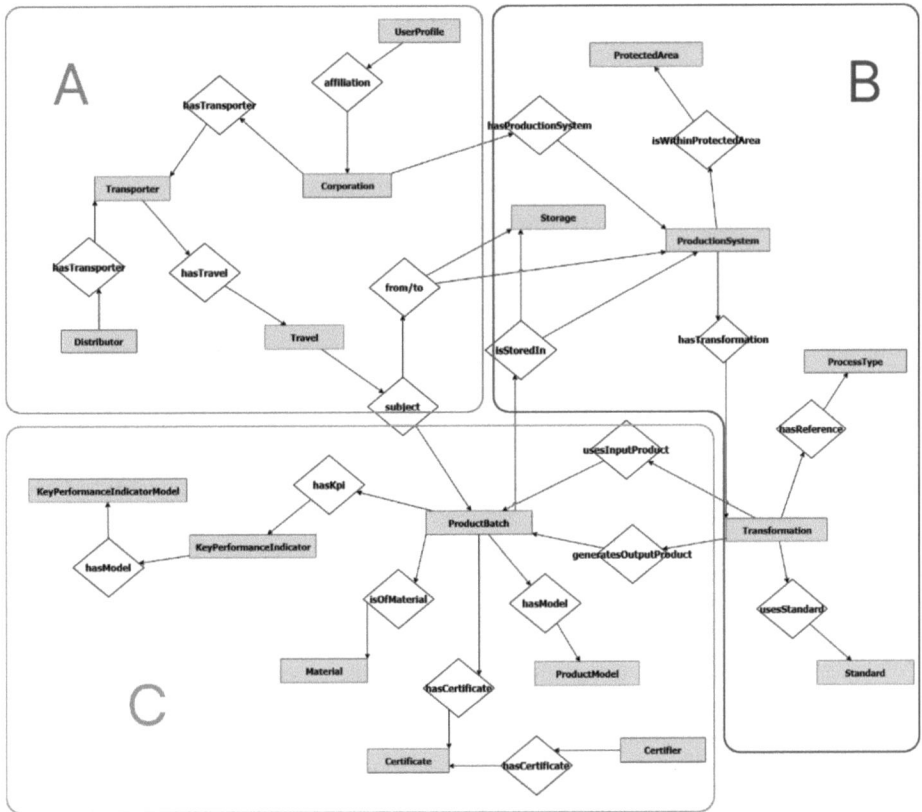

Fig. 1. Simplified context data model for supply chain tracking

in consistent travel routes with standardized transporters and products. Based on this assumption, rough estimates of Travels can be generated in terms of carbon footprint without the need to track every individual trip for each transporter.

A similar approach is applied to the Transformation entity. Considering the Transformation process as an archetype that remains largely consistent, with similar inputs, outputs, consumption, and requirements, similar to the Travel model, allows the BIT platform to generate rough estimates for carbon footprint, energy consumption, and other key requirements based on the defined ProductBatch archetype (Table 2).

The main objective is to ensure that all relevant information about bioproducts is transparently and readily accessible at any step of the supply chain. This approach empowers platform users with the data they need to make informed decisions, fostering efficiency and accountability within the process (Table 3).

Relationships between entities form a critical component of the NGSI-LD model, explicitly representing material flows, transformation processes, and organizational connections throughout the supply chain. Key relationships include "isStoredIn" (connecting products to their storage area), "generatesOutputProduct" and "usesInputProduct" (linking materials to transformation processes), and "hasProductionSystem" (connecting

Table 1. Description of the entity types of sector A.

Entity type	Description
User Profile	The User Profile entity carries information about the BIT user. Depending on his profile, they will be redirected to a different part of the application. He can be affiliated to a Corporation thanks to the "affiliation" relationship
Corporation	The Corporation entity regroups all kinds of economic actors: "Primary Producer", "Processor", "Manufacturer", "Waste Manager". At each step of the Product life cycle, a different Corporation can play a different role. An industry score can be calculated and later used in product environmental impact formulas
Distributor	The Distributor only has a distribution role in the Product life cycle. It only connects to its Transporters
Transporter	The Transporter is owned either by the Distributor or the Corporation. It can be a truck or a boat and holds its consumption information
Travel	When a Transporter moves a ProductBatch from a point A to a point B, a Travel entity is registered with the "subject" relationship pointing to the ProductBatch. The relationship also holds information about the starting points and the destinations, and the calculated route

Table 2. Description of the entity types of sector B.

Entity type	Description
ProductionSystem	The ProductionSystem is linked to a Corporation. It's the representation of a machine which applies a Transformation to a ProductBatch. It can also be used as a storage place for a ProductBatch
ProtectedArea	If a ProductionSystem entity point location is located within a ProtectedArea multipolygon the relationship between these two entities is automatically created
Transformation	The Transformation represents the effect of the ProductionSystem entity on the Product. It holds the average requirements in terms of energy, water, chemicals, etc., for a full transformation cycle. Its information is specific to the machine
Storage	Storage defines the way in which a ProductBatch is stored, from the mode of storage, volume of product stored, type of maintenance, carbon footprint derived from storage, etc.
Standard	The Standard allows us to know if a Transformation complies with any Standard
ProcessType	The ProcessType entity is a Reference for a Transformation in case information is missing. Less precise, it's a fallback for a Transformation in a given operation area

Table 3. Description of the entity types of sector C.

Entity type	Description
ProductBatch	The ProductBatch entity has a central place in the middle of the model. It connects with lots of other entities to know as much as possible about its processes and transportations
ProductModel	The ProductModel is used to carry repeated information of several ProductBatch and centralized generic product information
Material	The Material entity informs about whether the Product would be eligible to ISCC (International Sustainability and Carbon Certification)
Certifier	The Certifier entity was supposed to represent the user allowed to manage Certificates, but it will probably just be merged into the UserProfile entity in the near future
Certificate	The Certificate informs about the Product certifications level in terms of sustainability, circularity, social acceptability, etc.
KeyPerformanceIndicator	The KeyPerformanceIndicator holds the calculated value of a Circularity Indicator related to a specific ProductBatch
KeyPerformanceIndicatorModel	The KeyPerformanceIndicatorModel centralizes generic KPI information

physical machinery entities to organizational actors). These relationships enable traceability queries that can traverse the entire supply chain, from raw materials to finished products and beyond to end-of-life management.

4.2 The BIT Web Interface for Data Visualisation and Traceability in Sustainable Supply Chains

To confirm and evaluate the model performance and capabilities, a web interface has been developed on top of it, as illustrated in the Fig. 2 below. The BIT front-end application serves as the primary access point for users of the BioReCer platform. It seamlessly integrates interoperable data acquired through Apache NiFi [19], utilizing the Stellio Context Broker API as an intermediary. The application retrieves and presents data in various formats, depending on the user profile. It is important to note that the user interface presented here is just one of many display possibilities that the model allows.

The current NGSI-LD context data model, as presented here, enables the creation of an interface where users can search for any product stored in the database using keywords, geo-queries, or both, and display the results on a map. The native capabilities of the NGSI-LD API allow products to be visualized based on their proximity to the user's location. This functionality has significant potential for decision support, providing users with insights into the current locations of the products they are searching for.

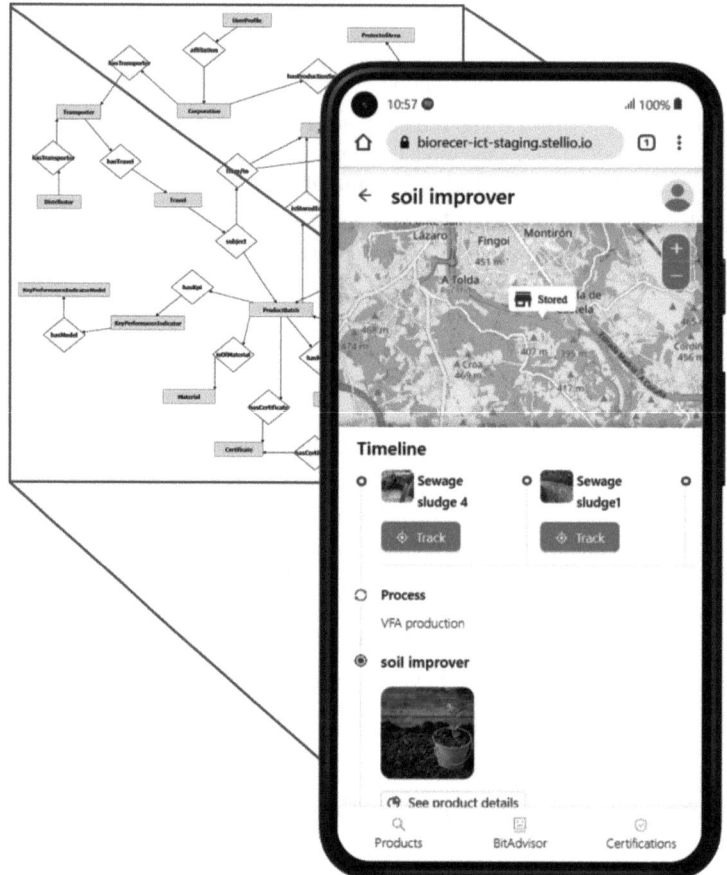

Fig. 2. The NGSI-LD data model enabling the BIT user interface

The current model also enables the exploration of the supply chain for any product within the system. By leveraging the known inputs and outputs of each Transformation entities, it is possible to trace the origins of a product, identifying the materials used as inputs, as well as the outputs generated at each stage. This structure allows navigation through the supply chain in both directions, providing visibility into key metrics such as energy consumption and other contextualized performance indicators at each step. Therefore, following the model structure of relationships, the application can calculate the energy consumption of every step and show a total value at any desired point of the supply chain.

Indeed, the capabilities of the model enable the automatic calculation of key performance indicators (KPIs) for a given product across the entire value chain. When the necessary data have been acquired and contextualized, the application can retrieve information on every Transformation and Travel that occurred upstream before the product's final state and or final product. This allows for the aggregation of carbon footprint, energy consumption, and other relevant metrics at each step to assess the overall environmental impact of production. This critical information is made available when needed or upon request by the user (Fig. 3).

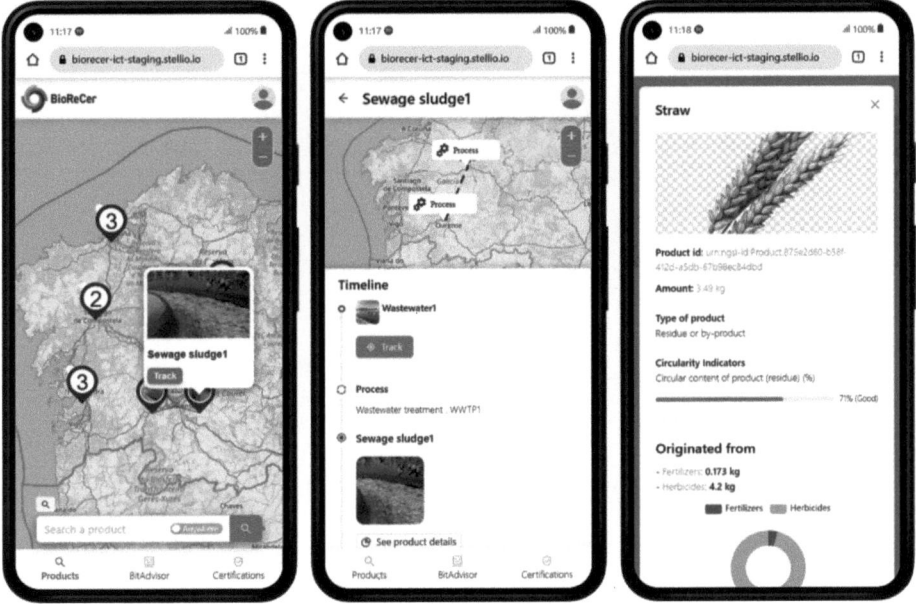

Fig. 3. The BIT user interface overview of supply chain exploration features

The BIT application implements role-based access control to provide appropriate functionality and data visibility for different user types, including primary producers, processors, certification bodies, and consumers. Each user type receives tailored configurations optimized for their specific decision-making needs. Of course, some features can be of interest for more than just one profile such as the supply chain exploration module. But some modules are more specifically oriented: for example, the module dedicated to the certification body user profile allows to browse standards and certification schemes by area of expertise and keywords and look for companies with high or low compliance scores.

4.3 An IoT Ready Architecture for Data Provisioning

The BIT platform implements a robust architecture which addresses the heterogeneity of data sources in bio-based supply chains through a flexible integration approach centered around Apache NIFI. The Apache NIFI component, responsible for data acquisition, validation, and transformation, can connect with any external data source. It serves as the primary entry point for integrating external information. For instance, NIFI can receive data from an IoT platform transmitting various sensor metrics, periodically retrieve data from Google Sheets, or query an external API. The component processes incoming data by applying sanitization, NGSI-LD payload transformations, and other necessary modifications before forwarding clean data to the Stellio Context Broker, ensuring compliance with the NGSI-LD interoperability standard and alignment with the shared context data model (Fig. 4).

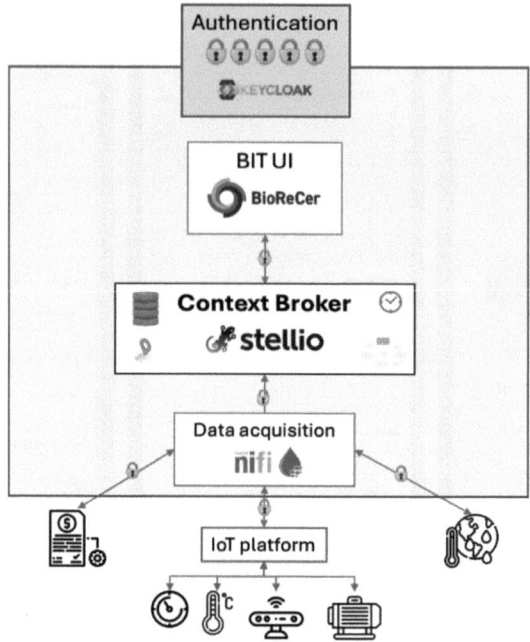

Fig. 4. The BIT general architecture secured with Keycloak

The BIT platform can handle automatically the registration of any devices relevant to bio-based supply chains, including environmental sensors, agricultural monitoring systems, process control sensors, and logistics tracking. The platform handles dynamically the device's communication protocol. Indeed, the architecture rely on an open source "device catalog" collectively maintained, and containing all known device information, including their respective decoding scripts and communication protocol details. For each registered device, the NIFI component can read the device catalog to decode and register incoming payloads into the Stellio context broker, on the proper target entities (e.g., Transformation, ProductBatch, ProductionSystem, Travel, etc.).

At the end of the chain, the BIT user interface consumes these data and displays them from the Stellio context broker the same way, whether the information comes from an IoT platform or any other source.

5 Case Studies and Validation

5.1 Case Studies Presentation

Four representative case studies across the European Union have been established to validate the project's assessment methodologies and certification schemes. These case studies represent diverse geographical distributions, feedstock types, and bio-based value chains.

Fishery and Sewage Sludge Value Chains (Spain)
The Galician case study focuses on the region's most relevant bio-based feedstocks: fishery/cannery waste and urban/industrial sewage sludge. The fishery sector, contributing approximately 5% to Galicia's GDP, generates valuable by-products including fish oil, partially hydrolyzed collagen, and seaweed derivatives rich in polysaccharides. These by-products can be utilized in multiple high-value applications including pharmaceuticals, cosmetics, biofertilizers, and food additives. Technical challenges include the development of efficient purification processes for fish oil and establishing transparent value chain traceability.

Urban Biowaste Valorization (Italy)
The Lombardy region case study examines the processing of the organic fraction of municipal solid waste (OFMSW), sewage sludge, agro-industrial waste, and water-based non-hazardous organic waste. These waste streams are converted into Volatile Fatty Acids (VFA), biopolymers, and biofertilizers through integrated biorefining processes. The primary technical challenges identified include the variable traceability of different waste streams, product quality consistency and market acceptance.

Agricultural Residue Valorization (Greece)
The Greek case study focuses on agricultural wastes and related by-product streams, including tree prunings, straw, crop residues, and other lignocellulosic biomass abundant in the region. These agricultural residues represent significant untapped resources that can be transformed into value-added products while addressing waste management challenges in rural areas. The technical challenges for this case study likely include the geographical dispersion of biomass resources, seasonal availability variations and standardization of heterogeneous feedstocks.

Forestry Waste Utilization (Sweden)
The Swedish case study investigates forestry waste streams including sawdust, bark, GROT (branches and treetops), and other wood processing residues. The forestry sector in Sweden generates substantial volumes of these by-products, which have significant potential for valorization in the bioeconomy. The primary technical challenges associated with this case study include optimizing logistics across large, forested areas and addressing seasonal harvesting patterns.

5.2 Tool Validation and Co-creation

The development of the BIT platform followed a collaborative co-creation approach that engaged diverse stakeholders throughout the design and implementation process. Stakeholders from Sweden and Italy were invited for user testing sessions of the web application to ensure usability, relevance, and efficiency when trying to reach their goals. Focus group discussions followed to refine ideas and functionalities. This methodology ensured that the resulting tool effectively addresses practical needs while accommodating various technical capabilities and operational constraints across the bio-based supply chain ecosystem.

Each case study was able to try and upload their supply chain data and successfully managed to explore and visualize them in a user-friendly manner. They were able to assess the overall carbon footprint, energy consumption, and other relevant metrics of the whole supply chain, but also at each step of the process for product batches and transformations.

6 Conclusion and Future Work

This paper has presented the BioReCer Information Tool (BIT), an NGSI-LD based approach for data visualization and traceability in sustainable supply chains and biological resources certification. The platform addresses critical challenges in the bio-based sector by implementing a comprehensive technical solution that bridges data interoperability gaps and enables transparent communication of circularity performance throughout complex value chains. Through the integration of IoT data sources, advanced context modeling, and intuitive visualization capabilities, the BIT platform provides a powerful tool for supporting certification processes and driving the transition toward circular economy practices in the bio-based sector.

The NGSI-LD context data model developed for BioReCer represents a significant advancement in the semantic representation of bio-based supply chains and their sustainability characteristics. By explicitly modeling entities, relationships, and temporal dimensions relevant to biological resources and their transformations, the data model provides a solid foundation for comprehensive traceability and circularity assessment. The validation of this model through diverse case studies confirms its applicability across different bio-based product categories and supply chain configurations.

The BIT web application successfully translates complex supply chain and certification data into accessible visualizations that support decision-making for various stakeholder groups. The modular design approach, with specialized components for supply chain visualization, circularity assessment, and certification management, ensures that the platform addresses the diverse needs of different users while maintaining consistent data representation and interaction patterns. The collaborative development process, involving stakeholders throughout the design and implementation phases, has resulted in a tool that effectively balances technical sophistication with practical usability.

The IoT-ready architecture, with Apache NIFI as a key intermediary component, provides a flexible approach to data provisioning from diverse sources. This architecture successfully addresses the heterogeneity of data formats, protocols, and quality levels typical of bio-based supply chains, ensuring reliable information flow from edge devices to the context broker. The implementation of comprehensive data quality management within the NIFI data flows maintains the integrity of the certification information base, a critical requirement for credible sustainability claims.

The case study implementations demonstrate the practical value of the BIT platform across different segments of the bio-based sector, from agricultural production through processing and end-of-life management. These implementations confirm the platform's ability to represent complex supply networks, track material transformations, and monitor sustainability metrics throughout product lifecycles. The positive feedback from stakeholders involved in these case studies validates the approach and suggests significant potential for broader adoption within the bio-based industry.

Building on the foundation established by the BioReCer Information Tool, several promising directions for future work have been identified that could further enhance the platform's capabilities and impact. These development pathways focus on emerging technologies and evolving regulatory frameworks that complement the NGSI-LD based approach presented in this paper.

The integration of Product Digital Passport frameworks represents a natural extension of the BIT platform's capabilities. As the European Union advances its Digital Product Passport initiative under the Circular Economy Action Plan, there is significant opportunity to align the BioReCer data model with these emerging standards. Future work will focus on extending the NGSI-LD model to incorporate the specific data points required for digital passports, developing standardized data exchange interfaces with passport repositories, and implementing user-friendly mechanisms for passport generation based on supply chain data already captured in the system.

Enhancing trust and security through blockchain integration presents another valuable direction for future development. While the current BIT platform implements robust data provenance tracking, the addition of blockchain technologies could provide immutable verification of critical certification data points, particularly at supply chain handoff points where trust gaps may exist. Future work will explore hybrid architectures that combine the contextual richness of NGSI-LD with the verification capabilities of distributed ledger technologies, focusing particularly on selective blockchain anchoring approaches that maintain system performance while providing cryptographic verification for key sustainability claims. This integration could significantly enhance the credibility of certification processes, particularly in complex international supply chains.

Finally, the application of predictive AI capabilities to sustainability forecasting represents a third promising direction. Building on the comprehensive historical data captured by the BIT platform, machine learning models could be developed to predict the environmental impacts of different supply chain configurations, optimize resource allocation for improved circularity, and identify emerging risks to sustainable practices. These predictive capabilities would transform the platform from a primarily descriptive tool to a prescriptive decision support system that actively guides users toward more sustainable choices.s

Acknowledgments. This research has been conducted as part of the BioReCer (Biological Resources Certifications Schemes) project, funded by the European Union's Horizon Europe research and innovation program under grant agreement No. 101060684. The authors gratefully acknowledge the financial support provided by the European Commission, which has made this work possible. We extend our sincere appreciation to all consortium partners of the BioReCer project for their valuable contributions, expertise, and collaborative spirit throughout the development and implementation of the BioReCer Information Tool. Special thanks are due to the case study partners who provided essential real-world testing environments and feedback that significantly improved the platform's functionality and usability. We would like to express our gratitude to all stakeholders who participated in the co-creation workshops, user testing sessions, and validation activities. Their insights, requirements, and practical feedback were instrumental in shaping the platform to address real-world needs in the bio-based sector.

References

1. BioReCer project. https://biorecer.eu/. Accessed 20 June 2025
2. IBM's Food Trust Overview. https://www.ibm.com/docs/en/food-trust?topic=overview. Accessed 20 June 2025
3. Walmart food traceability requirement page. https://one.walmart.com/content/food-safety/en_us/food-safety-requirements/food-traceability.html. Accessed 20 June 2025
4. Provenance. https://www.provenance.org/. Accessed 20 June 2025
5. VeChain. https://vechain.org/. Accessed 20 June 2025
6. Origin Trail, Supply chain page. https://origintrail.io/solutions/supply-chains. Accessed 20 June 2025
7. Tian, F.: A supply chain traceability system for food safety based on HACCP, blockchain & Internet of Things. In: 2017 International Conference on Service Systems and Service Management, Dalian, pp. 1–6 (2017)
8. Kouhizadeh, M., Saberi, S., Sarkis, J.: Blockchain technology and the sustainable supply chain: theoretically exploring adoption barriers. Int. J. Prod. Econ. **231** (2021)
9. Libelium smart agriculture page. https://www.libelium.com/libeliumworld/solution-success-stories/agriculture/. Accessed 20 June 2025
10. Bosch Environmental Sensors. https://www.bosch-sensortec.com/products/environmental-sensors/. Accessed 20 June 2025
11. Siemens, MindSphere page. https://www.siemens.com/fr/fr/produits/software/mindsphere.html. Accessed 20 June 2025
12. Microsoft Azure Digital Twins page. https://azure.microsoft.com/en-us/products/digital-twins. Accessed 20 June 2025
13. Boje, C., Guerriero, A., Kubicki, S., Rezgui, Y.: Towards a semantic construction digital twin: directions for future research. Autom. Constr. **114** (2020)
14. BIC. https://biconsortium.eu/. Accessed 20 June 2025
15. Wolfert, S., Ge, L., Verdouw, C., Bogaardt, M.-J.: Big data in smart farming – a review. Agric. Syst. **153**, 69–80 (2017)
16. El Bilali, H., Allahyari, M.S.: Transition towards sustainability in agriculture and food systems: role of information and communication technologies. Inf. Process. Agric. **5**(4), 456–464 (2018)
17. Wikipedia article. https://en.wikipedia.org/wiki/NGSI-LD. Accessed 20 June 2025
18. Stellio Context Broker Documentation. https://stellio.readthedocs.io/en/latest/. Accessed 20 June 2025
19. Apache NiFi. https://nifi.apache.org/. Accessed 20 June 2025

Open Access This chapter is licensed under the terms of the Creative Commons Attribution 4.0 International License (http://creativecommons.org/licenses/by/4.0/), which permits use, sharing, adaptation, distribution and reproduction in any medium or format, as long as you give appropriate credit to the original author(s) and the source, provide a link to the Creative Commons license and indicate if changes were made.

The images or other third party material in this chapter are included in the chapter's Creative Commons license, unless indicated otherwise in a credit line to the material. If material is not included in the chapter's Creative Commons license and your intended use is not permitted by statutory regulation or exceeds the permitted use, you will need to obtain permission directly from the copyright holder.

A Fair and Lightweight Consensus Algorithm for IoT

Sokratis Vavilis[✉], Harris Niavis, and Konstantinos Loupos

Inlecom Innovation, Athens, Greece
{sokratis.vavilis,harris.niavis,konstantinos.loupos}@inlecomsystems.com

Abstract. With the rapid growth of hyperconnected devices and decentralized data architectures, safeguarding Internet of Things (IoT) transactions is becoming increasingly challenging. Blockchain presents a promising solution, yet its effectiveness depends on the underlying consensus algorithm. Conventional mechanisms, such as Proof of Work and Proof of Stake, are often impractical for resource-constrained IoT environments. To address these limitations, this work introduces a fair and lightweight hybrid consensus algorithm tailored for IoT. The proposed approach minimizes resource demands on the nodes while providing a fair and secure agreement process. Specifically, it utilizes a distributed lottery mechanism to ensure fair block proposals without requiring dedicated hardware. In addition, to enhance trust and establish finality, a reputation-based voting mechanism is incorporated. Finally, we experimentally validated the key features of the proposed consensus algorithm.

Keywords: Blockchain · Internet of Things · IoT · Consensus algorithm

1 Introduction

As we enter an era of hyperconnected devices and decentralized data architectures, maintaining the integrity and security of transactions within such a distributed ecosystem presents significant challenges [26,35]. Factors such as device heterogeneity, insecure communication protocols, and limited computational resources create barriers to establishing trust in the IoT device lifecycle [44]. Integrating blockchain into IoT systems offers a transformative approach to enhance security, data integrity, and interoperability [11,24]. Especially in the last-mile delivery sector, where blockchain can enhance trust between urban logistics stakeholders through green smart contracts and digital identities [18], enabling transparent and tamper-evident coordination across the supply chain. At the core of this integration is the consensus algorithm, a fundamental component that dictates network throughput, transaction validity, and immutability. Developing efficient and resilient consensus mechanisms is essential to enable secure, scalable, and trustworthy decentralized IoT ecosystems.

Although fundamental consensus algorithms such as Proof of Work (PoW) and Proof of Stake (PoS) have been effective in general-purpose blockchains such as Bitcoin and Ethereum, they face unique challenges when applied to IoT ecosystems. The resource-intensive nature of PoW, for example, poses a significant hurdle in the resource-constrained environments characteristic of IoT devices [36,39,42]. Moreover, the need for energy-efficient consensus mechanisms becomes paramount as the proliferation of connected devices continues unabated.

This paper introduces a novel, fair and lightweight consensus algorithm tailored for the demands of Blockchain in IoT scenarios. The algorithm addresses limitations of existing mechanisms by optimizing for low resource consumption and fairness. It uses distributed lottery mechanisms to ensure equal participation and reputation-based block voting to foster trust in the process, all without depending on specific hardware, enabling its applicability to various devices.

This approach improves performance, security, and trust in IoT environments, such as decentralized last-mile delivery IoT networks, where secure and efficient sharing of logistics events between stakeholders is critical for traceability, non-repudiation, and regulatory compliance. By addressing key limitations of existing consensus mechanisms, the proposed solution can further promote broader blockchain adoption in real-world IoT and logistics applications.

The remainder of this paper is organized as follows. Section 2 introduces the reader to blockchain consensus algorithms by discussing related work. In Sect. 3 we present an overview of the proposed algorithm along with details of each phase, while Sect. 4 analyzes the experiments performed to assess the algorithm in terms of fairness and robustness to attacks. Finally, Sect. 5 concludes and points out directions for future work.

2 Related Work

Consensus algorithms are regarded as the cornerstone of blockchain as they provide the means for nodes to agree on the state of the blockchain. Specifically, they provide the mechanisms by which the nodes agree on the blocks to be added to the blockchain. The characteristics of the consensus algorithm are crucial for both the performance and the security properties of the blockchain solution. Various approaches have been proposed in the literature, originating from the field of distributed network systems and, more recently, from blockchain technology per se. Generally, consensus algorithms can be classified into two broad categories: voting-based and proof-based [20].

Traditional distributed systems approaches have been adapted for permissioned blockchains, as they share similar foundational assumptions. These methods rely on a voting mechanism, where nodes elect a leader and vote on block proposals, with final decisions determined by majority rule. Their primary goal is to ensure consistency and finality, guaranteeing that all nodes have a unified view of the blockchain state after each round. Nonetheless, these approaches tend to be more centralized and face scalability issues due to the high communication overhead required for consensus. Additionally, many voting-based models assume that most participating nodes are both honest and consistently available.

Over the past decades, Paxos [23] was the dominant voting-based consensus mechanism, though its inherent complexity hindered its adoption. To address this issue, Raft [31] was proposed as a more understandable alternative. Raft enhances clarity by introducing a straightforward leader election mechanism, in which each node has the chance to campaign for a Leader role for a given round. This is achieved by random time-based rounds during which a candidate solicits votes from their peers, requiring a majority to become the leader. Both approaches, although crash-fault tolerant, assumed the honesty (i.e., trust) of all participating nodes, restricting their applicability in many real-world scenarios. To this end, Byzantine fault-tolerant (BFT) solutions were suggested. One of the most widely applied and efficient mechanisms of this kind is the Practical Byzantine Fault Tolerance (pBFT) [9], which was designed to tolerate up to —(n-1)/3— malicious nodes. It uses a three-phase protocol, in which failing to reach consensus among all consensus nodes triggers the election of a new primary node (leader). The major drawback of pBFT is its quadratic complexity $O(n^2)$ compared to the linear complexity of the other two algorithms. Modern and scalable voting-based methods have also been proposed exclusively for blockchain networks. A prominent example is Proof of Vote [25], in which a committee of predefined members elects butler nodes, which are responsible for validating transactions and adding blocks to the chain.

Public permissionless blockchains face distinct security challenges, as they cannot assume trustworthy nodes and must prioritize performance due to their large scale. Their design emphasizes decentralization, security, and availability, often at the cost of immediate consistency and finality, leading to blockchain forks. To mitigate forks and achieve eventual consistency probabilistically, these blockchains employ mechanisms such as computational challenges (e.g., solving cryptographic puzzles) or leveraging node properties (e.g., stake ownership).

Regarding public blockchains, PoW [8,15] is the most popular consensus algorithm, as it is the one adopted by Bitcoin. PoW is a fully decentralized consensus mechanism that requires network members to expend computational effort to solve a mathematical puzzle. The first node to solve the puzzle adds a block to the chain. Verifying a solution to such a puzzle can be easily and undeniably confirmed by other nodes. PoW methods, however, experience low throughput and have a considerable environmental impact due to their high energy consumption. To overcome the latter, different mechanisms have been proposed in the literature, such as Proof of Capacity / Proof of Storage [4,16], which pose challenges related to data storage, which has a lower environmental footprint. Other approaches focus on physically selecting nodes employing the notion of randomness similar to participating in a lottery. For instance, in Proof of Elapsed Time [12] each node has a random timer, and the node that times out first adds a block to the chain. Similarly, in the Proof of Luck (PoL) algorithm [29], each node calculates a random number and the luckiest (e.g., smallest number) adds a block. To guarantee the honest behavior of the nodes and the indisputability of the random procedures, a Trusted Execution Environment (TEE) must be present in each node. Based on the above, it becomes apparent that these solutions require nodes to have a considerable number of resources or features, which renders them inappropriate for an IoT context [34,36,39,42]. Nonetheless, recent approaches like Proof of Verifiable Function (PoVF) utilize Verifiable Delay Functions (VDFs) and Verifiable Random Functions (VRFs) to achieve unpredictable node selection without relying on specialized hardware [43].

Apart from methods that depend on a physical challenge to select a node to add a block, other consensus algorithms make such a choice based on node properties. For instance, Proof of Authority [38] solutions rely on a set of preauthorized nodes whose identity and trust have already been verified. In each round, one node from that set is chosen to add a block, while the rest of the nodes in the network follow that decision. Although quite efficient, these approaches increase the level of centralization, and authorized nodes usually must be preapproved in the physical world [39,42]. The most prominent consensus algorithm in the aforementioned family is PoS [22], which is quickly gaining ground among modern cryptocurrencies. In this approach, the node to add a block is selected considering the stake (e.g., coins held) that a node has in the network. The underlying intuition is straightforward: the higher the stake, the greater the chances of being chosen to add a block. In this context, high-stakes nodes have an incentive for the network to operate as designed and be profitable. Moreover, nodes that do not fulfill their responsibilities may lose a portion of their stake (i.e., slashing) as punishment; thus, they are less likely to misbehave. PoS algorithms exist in many different flavors, such as Delegated PoS [3], which aim to be more efficient, or Pure PoS [13,19] solutions which aim to be more secure and fair. The main drawback of such approaches is that they are designed for cryptocurrencies where nodes have an economic incentive, making their application difficult in different contexts [39,42]. Lastly, such algorithms favor nodes with significant investments in the network, leading to increased centralization.

Although narrow in their applicability, PoS algorithms have been proven to be secure, efficient, and have limited resource requirements, thus gaining the interest of the research community. To overcome their applicability problems, researchers have worked on expanding their concept to cover a wider variety of properties. To this end, Proof of Importance (PoI) [30] algorithms have been proposed. This family of algorithms extends the concept of stake by considering other aspects that highlight the importance of a node in the network. For example, the contribution to the network, the number of successful transactions, and the trustworthiness of the nodes are considered. A particular type of PoI algorithm is Proof of Trust/Reputation algorithms [21,37,45] that assess the overall trustworthiness of nodes (e.g., based on their characteristics and past behavior) to select a node for adding a block to the chain. These methods show great potential, and there is substantial ongoing research in the field of IoT. However, the main challenge with these algorithms is related to their fairness, as nodes with higher importance tend to dominate the blockchain network.

In real-world applications, most consensus algorithms adopt a hybrid approach, combining different mechanisms to capitalize on their strengths while mitigating their shortcomings. One such example is Proof of Activity [6], which integrates elements of both PoW and PoS. In this model, a lightweight PoW process is used to mine a block, after which PoS is used for validation and block addition. Other notable hybrid approaches include Ethereum's PoS and Algorand [13,19], which, while primarily PoS-based, incorporate pBFT to enhance security and guarantee finality. Hybrid consensus mechanisms have also been explored in the IoT domain, such as PoEWAL [2], which combines PoW with PoL. Given their ability to balance security, efficiency, and scalability, hybrid approaches present a promising direction for IoT applications.

While we have already discussed the applicability of fundamental consensus paradigms in IoT, several recent studies proposed algorithms exclusively for IoT [34]. For example, [7] extends the HyperLedger Fabric framework with a novel approach to reduce the nodes participating in the consensus process, enhancing its suitability for IoT environments. In [14], a leader-based consensus algorithm is introduced, where leaders precompute a hash-based token (i.e., consensus code) and are assigned to manage transactions associated with it. This approach allows validators to create and commit blocks containing transactions under their supervision, thereby reducing the need for extensive inter-node coordination.

Lastly, two works share similarities with our algorithm: REVO [5] and LVCA [41]. REVO leverages reputation to form trusted committees that elect nodes for block addition, whereas LVCA employs a consensus based on random lottery voting among trusted nodes to select the block proposer. In contrast, our algorithm emphasizes fairness by randomly selecting block proposers from the global pool of nodes while using reputation only to validate and add blocks.

3 Fair and Lightweight Consensus

In this section, we discuss our consensus algorithm for IoT devices. First, we provide an overview of our proposal, and then we proceed to discuss the details of each phase of the algorithm.

3.1 Overview

Blockchain offers strong potential to address IoT security challenges due to its decentralized and immutable nature. However, the application of consensus algorithms in IoT remains a challenge. Resource-heavy mechanisms such as PoW are unsuitable for constrained devices, whereas lighter alternatives like PoS or PoI may unfairly favor nodes, effectively demotivating newcomers. Voting-based methods also struggle, as they often incur high communication overhead or lack resilience against malicious actors, especially in dynamic, public networks.

To address these limitations, we present a novel hybrid consensus algorithm for the IoT that is both lightweight and fair. As illustrated in Fig. 1, the algorithm comprises two main parts: a distributed lottery mechanism for block proposals and a reputation-based voting process to finalize the agreement on a specific block. The lottery mechanism employs Verifiable Random Functions (VRFs),

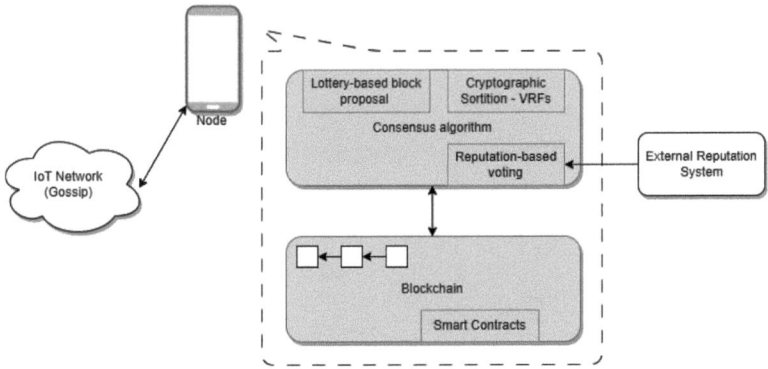

Fig. 1. Consensus algorithm structure overview

enabling nodes to generate a random lot in a verifiable and decentralized fashion. The voting process relies on a consortium of trusted nodes who vote for the best block proposal observed in a given round. The trust level of each node is determined through an external reputation system [1,17,40]. We note that our approach is not constrained to a particular reputation model; additionally, for the purposes of this work, we assume the existence of such a system.

Unlike other consensus algorithms, our approach suits resource-constrained environments by avoiding heavy computations and special hardware requirements like Trusted Execution Environments (TEE) as done, for instance, in PoL [29]. It ensures fairness, giving all nodes equal opportunity to propose blocks regardless of their resources. A trusted consortium of randomly selected high-reputation nodes vets only the best proposal per round without influencing block creation itself. Their incentive to preserve reputation for future committee and network participation discourages malicious behavior.

3.2 Algorithm Flow

The operational flow of the proposed algorithm is shown in Fig. 2. In particular, the nodes of the network compete with each other in adding a specific block to the chain by continuously executing the depicted process. At the beginning of each round (i.e., order of block to be added), nodes check whether they are in the correct round. This is achieved by assessing whether the node has received any messages regarding a round greater than the current height of the local blockchain known to the node. In case the node is out of sync with the other nodes, it requests from its neighboring nodes the most up-to-date chain available. Based on replies from its neighbors, the node adopts the longest valid chain (in terms of blocks). Otherwise, the nodes proceed with the block proposal.

During the block proposal phase, each node executes a cryptographic sortition algorithm to determine whether it is selected to propose a block. If chosen, the node generates a proposal using the randomly computed lot created in the

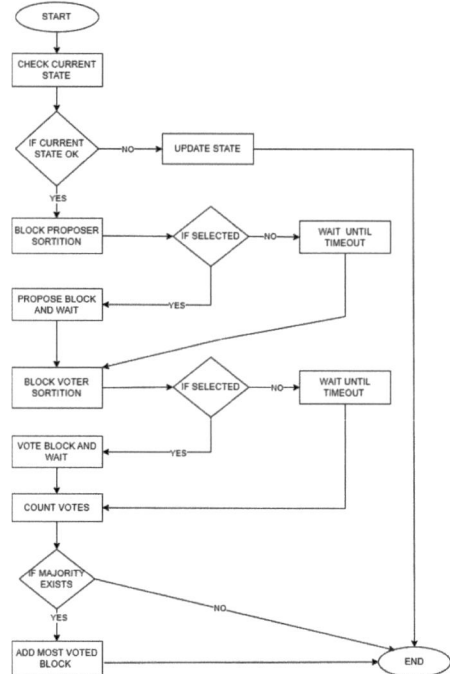

Fig. 2. Consensus algorithm execution flow

sortition process and broadcasts it. The node then waits for a predefined timeout period to allow message dissemination across the network. In the subsequent voting phase, the cryptographic sortition mechanism is applied again to select high-reputable nodes for the voting committee, which votes for the best block proposal (i.e., the one with the highest lot value) observed. After the voting timeout expires, each node determines whether a majority vote has been achieved for a proposed block, which is then added to the blockchain. Notably, if the number of participating nodes in each phase is known (e.g., in permissioned networks or through the reputation system), nodes can proceed to the next phase as soon as a majority is reached, without waiting for the full timeout. A reduced version of the algorithm, focused on permissioned networks but omitting sortition and dynamic timeouts, was previously implemented and validated in the Hyperledger Fabric platform [28].

During this process, each node executes in parallel a message handling routine, implementing a gossip protocol responsible for disseminating messages in the network. For efficiency reasons, previously seen messages and messages regarding rounds smaller than the currently known by a node are discarded. To protect the authenticity and integrity of the messages exchanged, they are cryptographically signed. We note that the communication approach used is

3.3 Algorithm Details

In the following, we describe the proposed consensus algorithm in detail.

Random Lotteries and Sortition. Random lotteries are central to our design, governing both block proposal and voting. Thus, we require a secure and verifiable method for generating randomness in a distributed setting. While a plethora of solutions exist [10,27,32,33], few suit resource-constrained IoT environments.

In our algorithm, we have chosen VRFs [27] to generate random lots, as it is a mature primitive with low computational and communication overhead. VRFs rely on public key cryptography to produce verifiable and uniform random numbers, in the form of hashes, given an input. In particular, a node can use a common public value as input to the VRF, known as the seed, and using its secret key will generate a random number, along with a cryptographic proof. Other nodes in the network can verify the validity of the produced lot using the accompanying proof, the seed used in its generation, and the public key of the lot issuer. Each round seed is calculated as $seed_r = VRF_{sk}(seed_{(r-1)}||role)$, where r represents a round. To properly utilize the properties of VRFs, the publicly known seeds used in each round must be chosen at random and not controlled by potential threat agents. To this end, we require that all agents' secret keys are defined well in advance the seed of a particular round (e.g., in the previous round or earlier). In this respect, we follow a similar strategy as the one discussed in [19], and we recommend the reader to read that work for more details.

Algorithm 1. Cryptographic Sortition

1: **procedure** SORTITION($sk, role, seed$)
2:     $lot, proof \leftarrow VRF_{sk}(seed||role)$
3:     **if** $role$ is $PROPOSAL$ **and** $lot \geq (THRESHOLD_{proposal} \times MAX\_LOT)$ **then**
4:         **return True**, $lot, proof$
5:     **else if** $role$ is $VOTE$ **and** $node\_reputation \geq THRESHOLD_{reputation}$ **and** $lot \geq (THRESHOLD_{vote} \times MAX\_LOT)$ **then**
6:         **return True**, $lot, proof$
7:     **else**
8:         **return False**
9:     **end if**
10: **end procedure**

As the number of participants in the network can be arbitrarily high, we need a mechanism to select random subsets of nodes to actively participate in the block proposal and voting process. This will effectively decrease the communication overhead introduced by exchanging a large number of block proposals or voting messages. To meet this requirement, we employ cryptographic sortition based on VRFs, to secure the random selection of nodes. Due to the use of VRFs, cryptographic sortition is reliable and verifiable by the nodes of the network. As shown in Algorithm 1, of a particular round and phase of the algorithm (i.e., block proposal or voting), a node calculates a random lot using the VRF and if this value is above a predefined threshold, then the node is selected for participation. Other nodes of the system can verify the outcomes of the sortition mechanism for a node using Algorithm 2. In the following, we discuss how the sortition mechanism is used for proposing blocks and voting.

Block Proposal. In each round, every node executes the sortition algorithm (Algorithm 1) to assess whether it is chosen to propose a block. Initially, the node calculates a random lot for the round using the VRF function along with its public key and the round's seed. If the calculated lot is over a specific threshold, the node proceeds to send its block, including the generated lot with the associated proof and the block's seed. The nodes receiving the block proposal employ Algorithm 2 to verify it. This algorithm verifies the VRF and then uses similar criteria to the ones used in the sortition. We note that the proposed sortition algorithms are designed to give equal chances of selection to every legible node, without prioritizing nodes based on any other factor than the random lot.

The threshold aims to limit the number of nodes proposing a block, hence limiting the number of proposals in the network. To calculate the threshold value, we use a weight (i.e., $THRESHOLD_{proposal}$) and multiply it by the maximum potential value of the VRF (i.e., MAX_LOT). Since VRF's output is uniformly distributed, the $THRESHOLD_{proposal}$ defines the anticipated ratio of nodes to be selected. For instance, by setting $THRESHOLD_{proposal} = 0.9$, we anticipate that only 10% of the nodes will be chosen.

We note that using this approach, the number of selected block proposers grows linearly to the number of nodes, although at a lower rate defined by $THRESHOLD_{proposal}$. However, in case the total number of nodes in the network is known, it is possible to dynamically adjust the $THRESHOLD_{proposal}$ value to fix the number of block proposers per round. Albeit trivial in a permissioned blockchain network, it is not the case for public permissionless networks. Nevertheless, assessing the potential number of nodes in a permissionless network is beyond the scope of the current work.

Finally, to limit the amount of information transmitted over the network, during the block proposal phase, only the generated lots along with their proofs and the proposed block's hash are sent over the network. Only after the end of the proposal phase is the actual winning block(s) transmitted. Next, we argue that due to the random selection procedure of our approach, potential adversaries have limited chances to be selected, which is relative to their ratio over the participants of the network. In practice, an adversary has to control over 50% of the network to have meaningful chances of being chosen and posing a threat in the operation of the blockchain. Such a scenario would be feasible in a permission-less network via Sybil-attacks since our algorithm lacks explicit protection against Sybil attacks. However, even in that case, attackers will only be able to slow down the rate of adding blocks to the chain. Such an effect could be mitigated by using the reputation system to detect patterns of malicious behavior and ban nodes that consistently misbehave. We leave the investigation of such an approach for future work. Lastly, we note that the safety of the blockchain would not be compromised due to the reputation-based voting mechanism.

Algorithm 2. Cryptographic Sortition Verification

1: **procedure** VERIFYSORTITION($pk, role, seed, lot, proof$)
2:     **if** $VerifyVRF_{pk}(lot, proof, seed\|role)$ is **False then**
3:         return **False**
4:     **end if**
5:     **if** $role$ is $PROPOSAL$ **and** $lot \geq (THRESHOLD_{proposal} \times MAX\_LOT)$ **then**
6:         return **True**
7:     **else if** $role$ is $VOTE$ **and** $node\_reputation \geq THRESHOLD_{reputation}$ **and** $lot \geq (THRESHOLD_{vote} \times MAX\_LOT)$ **then**
8:         return **True**
9:     **else**
10:        return **False**
11:     **end if**
12: **end procedure**

Voting Phase. After the block proposal phase, selected nodes of the network vote for the best block proposal, defined as the one with the highest lot value seen in the current round. To this end, every node of the network executes the sortition algorithm (Alg. 1) to decide whether they are part of the voting committee. The internal mechanics of the sortition algorithm for the voting phase are quite similar to the block proposal phase. In particular, VRFs are again used for the calculation of a random lot, however, there are a few notable differences.

First and foremost, we have introduced a reputation-based selection criterion. The intuition behind this choice is to only select nodes with a high enough reputation, that have already proven their good behavior in the network. To achieve this, a $THRESHOLD_{reputation}$ value must be set to a value that semantically denotes a high reputation. For instance, if a reputation system employs a $[0, 1]$ scale, the threshold value could be set to 0.8. We stress the fact that $THRESHOLD_{reputation}$ should be high enough, in order not to allow adversaries to be selected for voting. We also argue that with a high enough $THRESHOLD_{reputation}$ value, potential attackers would have to invest a significant amount of time and effort to achieve and retain a high enough reputation. Moreover, such a strategy should be done for multiple nodes, to achieve a high enough probability to be chosen by the sortition algorithm and form a majority.

Secondly, we introduce a different threshold ($THRESHOLD_{vote}$) for the lots used in the voter selection process. The reason for such a choice is that we want to adjust the size of voting committees in a different manner than for the block proposal. In addition, since the number of highly reputable nodes is known, this threshold value could be set to result in a voting committee of fixed size, further limiting the chances of potential adversaries being selected.

Block Addition. When the voting phase is completed, the network nodes will decide which block they will add to the blockchain. In particular, nodes first verify the received voting messages using the Algorithm 2. Then if a block has gathered the majority of the votes, it is added to the chain. Otherwise, the nodes do not add any blocks and proceed to the next round. It should be noted that nodes will only add blocks containing valid data (e.g., valid transactions). Assuming that the votes originate from high-reputable and trustworthy nodes, this approach ensures finality and safety of the chain. It should also be noted that due to the use of digitally signed messages and verifiable lots, the integrity of the exchanged information is not at stake. Therefore, potential adversaries may only be able to hinder the efficiency of the network by reducing the rate at which blocks are added to the chain.

4 Experiments and Discussion

This section presents the experiments conducted to evaluate the proposed consensus algorithm. We begin by outlining the objectives of the tests and the experimental setup, followed by a detailed discussion of the results obtained.

4.1 Experiment Objectives and Setting

Our primary objective is to evaluate the performance of the proposed consensus algorithm concerning its fairness property and its robustness to attackers. To this end, we have developed a reference implementation of the algorithm in Python. To enable communication and message exchange between the nodes, we implemented a REST API using Flask. Over this API a simple gossip protocol was developed, allowing nodes to spread messages to five random neighbors. Avoidance of repeating already sent messages, such as block proposals, was implemented to reduce communication overhead. It is important to note that in terms of performance, Python and Flash are sub-optimal choices. However, their ease of use enables rapid development which fits our objective of creating a Proof of Concept (PoC) to assess the algorithm's fundamental properties.

Regarding the configuration of the consensus algorithm, we made the following choices. Similarly to other algorithms [19,29], we set the overall round timeout period to 20 s (12 s for the block proposal and 8 s for voting). Empirically, this time window would be sufficient for spreading information (i.e., messages) over the network. Furthermore, to reduce the network overhead, we also set $THRESHOLD_{proposal} = 0.9$, allowing only 10% of the nodes to propose a block in each round. Regarding the voting mechanism, we set $THRESHOLD_{reputation} = 0.8$ (in the range [0,1]) to allow only high-reputable nodes to vote, while $Threshold_{vote}$ is set to allow a voting committee of 11 nodes. The nodes obtain reputation information via a reputation system. We also note that we set the size of the blocks at 200 kilobytes. Lastly, to reduce the computation overhead introduced by cryptographic operations (e.g., digital signatures of messages), we replaced such operations with less computationally intensive alternatives (e.g., time delays).

For the experiment, we deployed our sample implementation in the AWS cloud infrastructure, deploying up to 100 nodes (depending on the experiment use-case) in a similar number of t3.micro instances. Of these nodes, 15 are highly reputable, with a reputation of 0.9, from which the voting committee members will be randomly selected. Note that despite that the number of highly reputable nodes does not have an impact on the experiment results, it should be high enough to allow the formation of a voting committee. We assume that these nodes remain honest throughout the experiments. We also acknowledge that our experiment setting and implementation are not optimized for computational and network efficiency. However, we argue that this has a limited impact on our evaluation since we study the behavior of our algorithm with respect to fairness and robustness to adversaries rather than for efficiency and throughput.

4.2 Assessing the Fairness Property

A key aspect of our design is the fairness with which nodes add blocks to the chain. To evaluate this aspect, we measure the diversity of the block proposers (i.e., the number of different proposers) that have successfully added a block to the chain. We also measure the size of the globally accepted chain (i.e., the longest most common chain) and the number of nodes that are in sync with it.

We evaluate the fairness property with a set of experiments deploying different numbers of nodes. We let the consensus algorithm run for ten minutes for each number of nodes. We repeat this process five times and report the averages per metric. The results of our experiments are shown in Table 1.

Table 1. Fairness experiments results

Nodes	25	50	75	100
Blocks added	30	30	30	30
Proposer diversity	23	26	26	27
Nodes in sync	25	50	73	94

The first finding of our experiment is that the algorithm maintains a stable throughput of three blocks per minute, regardless of the number of nodes. Despite the relatively small-scale setup (up to 100 nodes) compared to large public blockchain networks with thousands of participants, the results indicate that the proposed approach performs as expected, with a round-time of approximately

20 s. Furthermore, most nodes stay synchronized with the global chain, though a slight drop in synchronization is observed as the network size increases. After inspecting the node logs, we found that the desynchronized nodes were lagging for a block, which could be the result of network discrepancies. These nodes are expected to catch up with the last block during the next round. Lastly, regarding the fairness property of our consensus algorithm, we notice that the block proposer diversity grows from 76% to 90% as the number of nodes increases. This behavior is expected because the uniform distribution of the probability of the lotteries is better expressed on a greater scale.

4.3 Robustness to Attacks

Before we move on to discuss how we assess the robustness of attacks, we need to define the attacker's profile. In this regard, we consider that an attacker can deploy multiple malicious nodes that will not respect the consensus protocol. Since attackers cannot influence the integrity of the exchanged messages and cannot forge the random lots (due to the use of VRFs), they concentrate their efforts on selecting which information to disseminate over the network to affect the block added to the chain. Malicious nodes may selectively forward block proposals and votes aiming to stop the expansion of the chain with new blocks. Next to that, malicious nodes aim to increase the chances that a malicious node is chosen to propose a block, effectively halting the addition of a block for that round. In our experiments, we assume that malicious nodes cannot have a high reputation and thus cannot participate in the voting committee.

To assess the robustness of our algorithm against attackers, we employ metrics similar to those used in the previous section. In particular, we measure how the presence of attackers impacts the fairness and expansion of the chain. We do this with a set of experiments deploying 100 nodes and a varying ratio of malicious nodes on the network. For each case, we let the consensus algorithm run for ten minutes. We note that by definition, the malicious nodes are deliberately not in sync with the chain. The results of our experiments are shown in Table 2

Table 2. Robustness to attacks results

Malicious Nodes	0	10	20	30	40	50
Blocks added		30	26	26	25	18	15
Proposer diversity	27	23	23	22	17	14
Nodes in sync		94	84	75	68	58	47

Evaluating the results, we observe that the ratio of malicious nodes has an impact on the throughput of the blockchain network. In particular, we notice a small drop of 14% in the number of blocks added when the ratio of attackers is relatively low (10% to 30%), however, a more significant drop of 40% -50% is noticed when the number of malicious nodes represents almost half of the nodes.

This behavior can be justified by the fact that the ratio of malicious nodes in the network is proportional to the probability of being chosen to propose a block. Such an effect could be mitigated by using the reputation system to decrease the reputation of nodes that consistently misbehave, and eventually ban them from the network. We leave the investigation of this use case for future work. We note that regardless of the drop in overall throughput, the network manages to operate with both safety and finality. Last but not least, it is noticed that the block proposer diversity (i.e., fairness) and the (honest) nodes synchronized to the global chain are not affected by the presence of adversaries in the network. This is evident in the scenario with 50 malicious nodes, where 14 of 15 blocks were proposed by different nodes, and 47 out of 50 nodes remained synchronized.

5 Conclusion & Future Work

This work introduced a fair and lightweight hybrid consensus algorithm designed for IoT environments, balancing security, fairness, and efficiency. It combines random lotteries for fairness with reputation-based voting for safety and finality, all with minimal resource demands. A sample implementation validated the algorithm's fairness property and its resilience against adversarial attacks. However, real-world efficiency and scalability in IoT remain open for further investigation. Finally, future research could further leverage the reputation mechanism to improve security and performance without compromising fairness.

Acknowledgment. This work has received funding from the EU Horizon Europe Programme in the framework of the "Upscaling Innovative Green Urban Logistics Solutions Through MultiActor Collaboration and PI-Inspired Last Mile Deliveries" project (URBANE), under GA Number 101069782

References

1. Aaqib, M., Ali, A., Chen, L., Nibouche, O.: IoT trust and reputation: a survey and taxonomy. J. Cloud Comput. **12**(1), 1–20 (2023)
2. Andola, N., et al.: PoEWAL: A lightweight consensus mechanism for blockchain in IoT. Pervasive and Mobile Computing **69** (2020)
3. Antheman, D.: DPOS consensus algorithm - the missing white paper (2016). https://steemit.com/dpos/@dantheman/dpos-consensus-algorithm-this-missing-white-paper
4. Ateniese, G., Bonacina, I., Faonio, A., Galesi, N.: Proofs of space: when space is of the essence. In: Abdalla, M., De Prisco, R. (eds.) SCN 2014. LNCS, vol. 8642, pp. 538–557. Springer, Cham (2014). https://doi.org/10.1007/978-3-319-10879-7_31
5. Barke, S., Srivastava, G.: ReVo: a hybrid consensus protocol for blockchain in the internet of things through reputation and voting mechanisms. In: 2024 IEEE 21st Consumer Communications & Networking Conference (CCNC), pp. 1–8. IEEE (2024)

6. Bentov, I., Lee, C., Mizrahi, A., Rosenfeld, M.: Proof of activity: extending bitcoin's proof of work via proof of stake. ACM SIGMETRICS Performance Evaluation Rev. **42**(3), 34–37 (2014)
7. Biswas, S., et al.: PoBT: a lightweight consensus algorithm for scalable IoT business blockchain. IEEE Internet Things J. **7**(3), 2343–2355 (2019)
8. Bitcoin, N.S.: Bitcoin: A peer-to-peer electronic cash system (2008)
9. Castro, M., Liskov, B.: Practical byzantine fault tolerance and proactive recovery. ACM Trans. Comput. Syst. (TOCS) **20**(4), 398–461 (2002)
10. Choi, K., Manoj, A., Bonneau, J.: SoK: distributed randomness beacons. Cryptology ePrint Archive (2023)
11. Christidis, K., Devetsikiotis, M.: Blockchains and smart contracts for the Internet of Things. IEEE Access **4**, 2292–2303 (2016)
12. Coorporation, I.: Intel Sawtooth lake (2016). https://web.archive.org/web/20161025232205/https://intelledger.github.io/introduction.html
13. Dimitri, N.: Proof-of-Stake in algorand. Distrib. Ledger Technol. **1**(2), December 2022
14. Dorri, A., Jurdak, R.: Tree-chain: A lightweight consensus algorithm for iot-based blockchains. In: 2021 IEEE International Conference on Blockchain and Cryptocurrency (ICBC), pp. 1–9. IEEE (2021)
15. Dwork, C., Naor, M.: Pricing via processing or combatting junk mail. In: Brickell, E.F. (ed.) CRYPTO 1992. LNCS, vol. 740, pp. 139–147. Springer, Heidelberg (1993). https://doi.org/10.1007/3-540-48071-4_10
16. Dziembowski, S., Faust, S., Kolmogorov, V., Pietrzak, K.: Proofs of space. In: Annual Cryptology Conference, pp. 585–605. Springer (2015)
17. Fortino, G., Fotia, L., Messina, F., Rosaci, D., Sarné, G.M.: Trust and reputation in the internet of things: state-of-the-art and research challenges. IEEE Access **8**, 60117–60125 (2020)
18. Franklin, R., et al.: The urbane innovation transferability platform: Learnings for decarbonising last-mile delivery networks. In: Climate Crisis and Resilient Transportation Systems, pp. 389–400. Springer Nature Switzerland (2025)
19. Gilad, Y., Hemo, R., Micali, S., Vlachos, G., Zeldovich, N.: Algorand: scaling byzantine agreements for cryptocurrencies. In: Proceedings of the 26th Symposium on Operating Systems Principles, pp. 51–68 (2017)
20. Jain, A.K., Gupta, N., Gupta, B.B.: A survey on scalable consensus algorithms for blockchain technology. Cyber Secur. Appl. **3**, 100065 (2025)
21. Kang, J., et al.: Toward secure blockchain-enabled internet of vehicles: optimizing consensus management using reputation and contract theory. IEEE Trans. Veh. Technol. **68**(3), 2906–2920 (2019)
22. King, S., Nadal, S.: Ppcoin: Peer-to-peer crypto-currency with proof-of-stake. self-published paper (2012)
23. Lamport, L.: The part-time parliament. ACM Trans. Comput. Syst. **16**(2), 133–169 (1998)
24. Lee, I.: Internet of Things (IoT) cybersecurity: Literature review and IoT cyber risk management. Future Internet **12**(9) (2020)
25. Li, K., Li, H., Hou, H., Li, K., Chen, Y.: Proof of vote: a high-performance consensus protocol based on vote mechanism & consortium blockchain. In: 2017 IEEE 19th International Conference on High Performance Computing and Communications; IEEE 15th International Conference on Smart City; IEEE 3rd International Conference on Data Science and Systems, pp. 466–473 (2017)
26. Lu, Y., Xu, L.D.: Internet of Things (IoT) cybersecurity research: A review of current research topics. IEEE Internet Things J. **6**(2), 2103–2115 (2019)

27. Micali, S., Rabin, M., Vadhan, S.: Verifiable random functions. In: 40th annual symposium on foundations of computer science (cat. No. 99CB37039), pp. 120–130. IEEE (1999)
28. Michalopoulos, F., Vavilis, S., Niavis, H., Loupos, K.: Integrating a hybrid lightweight consensus algorithm in hyperledger fabric. In: International Summit on the Global Internet of Things and Edge Computing, pp. 188–203. Springer (2024)
29. Milutinovic, M., He, W., Wu, H., Kanwal, M.: Proof of luck: an efficient blockchain consensus protocol. In: Proceedings of the 1st Workshop on System Software for Trusted Execution, pp. 1–6 (2016)
30. Niavis, H., Loupos, K.: Consenseiot: a consensus algorithm for secure and scalable blockchain in the IoT context. In: Proceedings of the 17th International Conference on Availability, Reliability and Security, pp. 1–6 (2022)
31. Ongaro, D., Ousterhout, J.: In search of an understandable consensus algorithm. In: 2014 USENIX Annual Technical Conference (USENIX ATC 14), pp. 305–319 (2014)
32. Raikwar, M., Gligoroski, D.: Sok: decentralized randomness beacon protocols. In: Australasian Conference on Information Security and Privacy, pp. 420–446. Springer (2022)
33. Randao.org: Randao: Verifiable random number generation (2017). https://www.randao.org/whitepaper/Randao_v0.85_en.pdf
34. Sahraoui, S., Bachir, A.: Lightweight consensus mechanisms in the internet of blockchained things: Thorough analysis and research directions. Digital Communications and Networks (2025)
35. Sicari, S., Rizzardi, A., Grieco, L., Coen-Porisini, A.: Security, privacy and trust in Internet of Things: the road ahead. Comput. Netw. **76**, 146–164 (2015)
36. Stefanescu, D., Montalvillo, L., Galán-García, P., Unzilla, J., Urbieta, A.: A systematic literature review of lightweight blockchain for IoT. IEEE Access (2022)
37. Sun, L., Yang, Q., Chen, X., Chen, Z.: RC-chain: reputation-based crowdsourcing blockchain for vehicular networks. J. Netw. Comput. Appl. **176**, 102956 (2021)
38. Szilágyi, P.: Eip-225: Clique proof-of-authority consensus protocol (2017). https://eips.ethereum.org/EIPS/eip-225
39. Tripathi, G., Ahad, M.A., Casalino, G.: A comprehensive review of blockchain technology: underlying principles and historical background with future challenges. Decision Anal. J. **9**, 100344 (2023)
40. Vavilis, S., Petković, M., Zannone, N.: A reference model for reputation systems. Decis. Support Syst. **61**, 147–154 (2014)
41. Verma, S., Chandra, G., Yadav, D.: LVCA: an efficient voting-based consensus algorithm in private Blockchain for enhancing data security. Peer-to-Peer Networking Appl. **18**(2), 1–19 (2025)
42. Wen, Y., Lu, F., Liu, Y., Cong, P., Huang, X.: Blockchain consensus mechanisms and their applications in IoT: a literature survey. In: Algorithms and Architectures for Parallel Processing: 20th International Conference 2020, NY, USA, pp. 564–579. Springer (2020)
43. Xiong, C., Yang, T., Wang, Y., Dong, B.: PoVF: Empowering decentralized blockchain systems with verifiable function consensus. arXiv preprint arXiv:2501.01146 (2025)

44. Yousefnezhad, N., Malhi, A., Främling, K.: Security in product lifecycle of IoT devices: a survey. J. Netw. Comput. Appl. **171**, 102779 (2020)
45. Zou, J., et al.: A proof-of-trust consensus protocol for enhancing accountability in crowdsourcing services. IEEE Trans. Serv. Comput. **12**(3), 429–445 (2018)

Open Access This chapter is licensed under the terms of the Creative Commons Attribution 4.0 International License (http://creativecommons.org/licenses/by/4.0/), which permits use, sharing, adaptation, distribution and reproduction in any medium or format, as long as you give appropriate credit to the original author(s) and the source, provide a link to the Creative Commons license and indicate if changes were made.

The images or other third party material in this chapter are included in the chapter's Creative Commons license, unless indicated otherwise in a credit line to the material. If material is not included in the chapter's Creative Commons license and your intended use is not permitted by statutory regulation or exceeds the permitted use, you will need to obtain permission directly from the copyright holder.

Author Index

B
Beliatis, Michail J.   175
Bertozzi, Nicolò   141

C
Cabanillas, Lucía   93
Campos, Enrique Mármol   57
Cenholt, Lasse   188
Chauhan, Ritu   39, 159
Cortés-Delgado, Francisco J.   57

D
Djenouri, Djamel   57

E
Esbrí, Miguel Ángel   93

F
Ferrera, Enrico   141

G
García, Natalia Borgoñós   125
Geraci, Anna   141
Gómez, Antonio Fernando Skarmeta   125

H
Hernández-Ramos, José L.   57

J
Jackman, Jack   3
Jena, Mehak   39
Julian, Matilde   93

K
Krenn, Stephan   57

L
Lacalle, Ignacio   93
Latif, Shahid   57
Lauterbach, Thomas   18
Le Gall, Franck   206

Leniston, Darren   3
Loupos, Konstantinos   224

M
Magnani, Romain   206
Malik, Ammar   3
Mishra, Aarushi   159

N
Niavis, Harris   224

O
O'Donnell, Terence   3

P
Padilla, María Hernández   125
Palau, Carlos E.   93
Pastrone, Claudio   141
Pérez, Jose Vivo   125
Presser, Mirko   71, 188
Puiu, Andrei   57

R
Raggett, Dave   111
Robert, Joerg   18
Ryan, David   3

S
Sacchet, Marco   141
Singh, Dhananjay   39, 159
Singh, Parwinder   71
Skarmeta, Antonio   57

T
Tsitsiva, Christina   175

U
Ul Haq, Asim   71

V
Vaño, Rafael   93
Vavilis, Sokratis   224
Vizitiu, Anamaria   57

If you have any concerns about our products,
you can contact us on
ProductSafety@springernature.com

In case Publisher is established outside the EU,
the EU authorized representative is:
**Springer Nature Customer Service Center GmbH
Europaplatz 3, 69115 Heidelberg, Germany**

Printed by Libri Plureos GmbH
in Hamburg, Germany